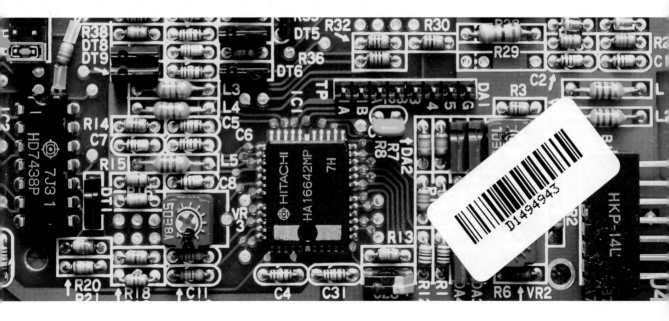

CONTEMPORARY CASE STUDIES

Economy &
Development

David Burtenshaw

Series Editor: Sue Warn

philip allan
UPDATES

For Amelie and Matteo, the new generation

Philip Allan Updates, part of the Hodder Education Group, an Hachette Livre UK company, Market Place, Deddington, Oxfordshire OX15 0SE

Orders

Bookpoint Ltd, 130 Milton Park, Abingdon, Oxfordshire OX14 4SB
tel: 01235 827720
fax: 01235 400454
e-mail: uk.orders@bookpoint.co.uk

Lines are open 9.00 a.m.–5.00 p.m., Monday to Saturday, with a 24-hour message answering service. You can also order through the Philip Allan Updates website: www.philipallan.co.uk

© Philip Allan Updates 2006

ISBN 978-1-84489-202-0

Front cover photograph reproduced by permission of Szasz-Fabian Jozsef/sxc.hu

Printed in Singapore

Philip Allan Updates' policy is to use papers that are natural, renewable and recyclable products and made from wood grown in sustainable forests. The logging and manufacturing processes are expected to conform to the environmental regulations of the country of origin.

P01108

Contents

Introduction

Economic geography is the study of where and how people make their living. It analyses how in different countries the sectors of the economy are interdependent (Figure 1)

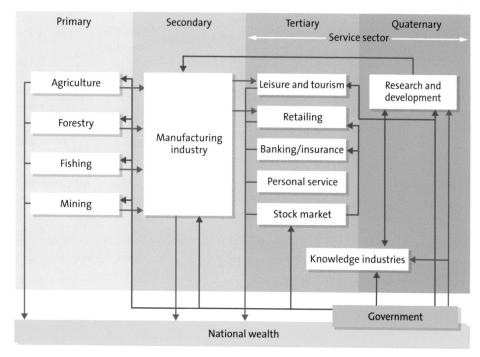

Figure 1
The modern economy

At a global scale, economic geography looks at how countries relate to each other through investment, trade and aid, and how these interactions have both positive and negative consequences for people and their environments (Figure 2).

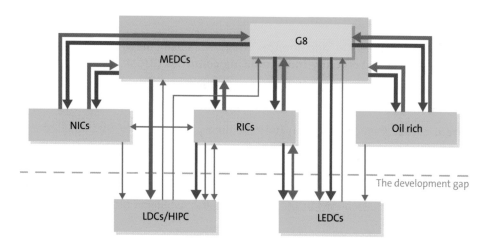

Figure 2 The links between different types of economy

About this book

The world is made up of 243 countries, each with its own economy, economic system and economic problems. This structure is outlined in Part 1.

The world is an unequal place. Part 2 takes a brief look at the development indicators and the changes that have taken place in geographers' understanding of global inequalities. It is not just a world of MEDCs and LEDCs, but a much more complex web of interdependent economies. While some countries prosper, others are plunged into increasing debt. The starkest manifestation of this divide is the development gap (Figure 2).

The main theme of recent change is that of globalisation and global shift of economic activity, which is the focus of Parts 3 and 4. The growth of transnational companies (TNCs) is illustrated by a range of modern firms that generate the products that we use today. There is a downside to globalisation, which is exemplified by the case of Saarland. The upside is illustrated by the NICs, and by the spectacular rise of India and China.

Part 5 provides case studies of the changing location of a major basic industry, steel, and of the process of locational choice of one manufacturer, Rolls-Royce. These show how site selection is subject to behavioural, organisational and political factors.

Services are an important component in the modern economy. Part 6 examines some of the principles underpinning the location of office services and returns to the theme of globalisation with an example of offshoring.

Part 7 uses case studies to examine a range of policies that have attempted to reduce economic inequalities: trade, Fair Trade, Foreign Direct Investment and aid, in its many forms, while Part 8 looks ahead to the future of work, the rise of China and the growth of green tourism.

Finally, Part 8 gives advice on answering a range of questions that involve stimulus material and case studies.

Key terms

Asian tigers: journalistic title for the Asian NICs (see **newly industrialising countries (NICs)** and **recently industrialising countries (RICs)**).

Behavioural factors: those factors relating to the way people react to, and modify, their environment as a result of their own perception, knowledge and opinions.

Bid rent curve: a graph based on bid rent theory, which states that land values decrease with increasing distance from a centre. It is normally applied to urban land values.

BPO: business process offshoring or outsourcing. This involves the movement of services to low-cost locations in LEDCs, NICs or former Soviet bloc states.

Break of bulk: a location for industry at the point where bulk cargoes are unloaded — normally a port location.

Capital intensive industry: manufacturing that depends heavily on investments in machinery and automation to produce goods. Car assembly today, using sophisticated robots, is a case in point.

Chaebol: Korean word for a highly centralised holding company, usually founded and run by a single person.

Colonialism: the period when the European powers ruled over many areas of the world. The colonies were dependent on the colonial powers, who exploited their resources and cheap labour (see **neocolonialism**).

Deindustrialisation: the decline of regionally important manufacturing industries.

Dependent economy: an economy that relies for investment on outside sources and/or state funding.

Development gap: the gap in levels of development and economic and social well-being between the economically advanced countries and the rest of the world. It has almost replaced the Brandt north–south divide as a depiction of global inequality.

Economies of scale: the gains that can be made from large-scale production. Unit costs are reduced.

Export-oriented industrialisation: manufacturing mainly for export rather than for local consumption.

Fair Trade: trade involving goods and services where the purchaser knows who produced them and who benefits from their trade. It began as a movement to benefit mainly foodstuff producers in the tropics, when big producers were driving small-scale producers out by cutting the price paid for produce.

FDI: foreign direct investment.

Fragile states: those countries where the government cannot or will not deliver core aspects of life to the population. Many are war zones.

G8: the world's most developed and economically influential countries (USA, Japan, UK, Germany, Canada, France, Italy and Russia).

Global companies: companies operating in different countries but with all units operating to a single plan. They are also referred to as transnational companies (**TNCs**).

Global interdependence: the way in which economies and societies are interlinked.

Globalisation: the economic interdependence between the leading nations in trade, investment and cooperative commercial relationships, and in which there are relatively few artificial restrictions on the cross-border movement of people, assets, goods or services. It is one of the aspects of internationalisation. Globalisation is the generic term for the process of integration in the realms of trade, economic relations and finance, but also includes social relations, knowledge, culture and politics). It has been aided by the ICT revolution that has reduced distance and, indeed, time.

Global shift: the movement of economic production in the recent past to new, cheaper locations associated with the rise of Japan and the NICs.

Gross domestic product (GDP): the value added by all resident production of goods and services plus taxes, but excluding overseas earnings and profits that go to foreign investors.

Gross national income (GNI): the sum of all value added by production of goods and services in a country plus taxes and income from abroad.

Gross national product (GNP): GDP plus income earned from overseas investment.

HDI: human development index. This UN measure is based on income per capita, adult literacy and life expectancy.

HIPC: heavily indebted poor country. A term developed by the World Bank.

Import substitution: activity designed to replace imports with locally produced goods made under licence. It is often an early stage in **NIC/RIC** growth.

Inertia: industries remaining in their original location well after the original location factors have gone.

Internationalisation: the spread of economic processes across boundaries to create a more extensive pattern of economic activity.

Iron curtain: the boundary between capitalist western Europe and the former Soviet bloc, which divided Europe between 1945 and 1989.

Keynesian interventionism: post-1930s perceived wisdom that economic development could be assisted by state intervention and control of aspects of the economy. The welfare state and regional development policy in the UK were based on this concept.

Labour-intensive industry: manufacturing dependent on a large labour force. Car assembly 50 years ago used much labour on production lines.

Labour pool: the available labour in an area. It might be measured in terms of people with a particular set of skills or the unemployed in search of employment.

LEDCs: less economically developed countries.

LDCs: least developed countries.

LICUS: low-income country under stress.

Location quotient: a statistic that measures the degree of concentration of an economic activity in an area. It enables agglomeration to be recognised.

MEDCs: more economically developed countries.

Mercantilism: seventeenth- and eighteenth-century increase in wealth, based on a positive overseas trade balance.

Multinational company (MNC): a number of separate, disparate subsidiary units operating on a semi-independent basis.

Multiplier effect: the range of effects of a decision to invest or locate in a region that lead to further direct, indirect, intended and unintended consequences. It can include the positive effects of new employment on jobs in linked and other activities, and the effects on regional income.

Neocolonialism: the continuing domination of the economies of former colonies by companies from the colonial power. Even aid can be neocolonial, for example the UK preferring its former colonies.

Newly industrialising countries (NICs): a term first coined by the World Bank to identify eight countries whose growth was due to rapid industrialisation — Malaysia, Hong Kong, Taiwan, Singapore, South Korea (see **Asian tigers**), China, Brazil and Mexico (see **RICs**).

Newly industrialising economies (NIE): an alternative term used for **NICs** and **RICs**.

Office decentralisation: the process of moving office jobs from major metropolitan centres to other locations in the suburbs and other towns and cities in the country. It was very evident between 1964 and 1979 in the UK.

Offshore banking: financial activities in countries where the regulations enable companies to avoid certain taxes, for example in Guernsey, the Isle of Man, the Cayman Islands and Labuan (Malaysia).

OPEC: Organisation of Petroleum Exporting Countries, with its HQ in Vienna. It represents mainly Middle Eastern oil producers.

Post-industrial society: society where a well-paid professional elite is served by a low-paid service sector.

Primary sector: the group of activities that includes agriculture, forestry, fishing and mining. It involves the collection and use of natural resources.

Product life cycle: the five stages in the cycle of a manufactured product, beginning with innovation, followed by growth of output, maturity of production at the peak, saturation of the market and obsolescence as new products replace the old ones.

Protectionism: a policy adopted by countries and trade blocs to give advantage to their own producers by excluding the products of other countries.

Quaternary sector: the skilled service sector, including government, research and development, administration, banking, insurance and financial management. It involves the assembly, transmission and processing of information and knowledge.

Rationalisation: the process of achieving greater efficiency and profit in an organisation. It often follows mergers and competition from other producers overseas, and may lead to deindustrialisation and the loss of jobs.

Raw material location: industrial production located at the source of the major raw material(s).

Recently industrialising countries (RICs): a term that acknowledges the growth of the economies of countries since 1990 and which refers both to **NICs** and countries such as India and China.

Reindustrialisation: the creation of new industries to replace employment in old industrial regions.

Secondary sector: manufacturing and the processing of raw materials.

Service sector: a general term used to cover a range of tertiary and quaternary economic activities.

Soviet bloc: those countries allied to the USSR in the period 1945–89.

Structural adjustment programme (SAP): imposed strict conditions on countries receiving aid. Receivers were obliged to cut back on healthcare, education, sanitation and housing programmes.

Structural unemployment: unemployment caused by people having the wrong skills to match the new jobs in an area.

Tertiary sector: the sector that enables goods to be traded. It includes wholesaling, retailing, transport and entertainment, including tourism and personal services.

Trade: selling and buying of the resources for, and products of, the four sectors of the economy.

Trade bloc: regionally based international organisations that have been established to encourage trade between members and to protect economic activity within the bloc by creating tariff walls. The EU, LAFTA and ASEAN are examples.

Transnational company: see **Global company**.

UNDP: United Nations Development Programme.

WTO: World Trade Organization.

Setting the scene

Economic geography is concerned with the distribution of economic activities, the factors that explain why such activities take place in these locations, the processes that lead to changing distributions and the changing relative importance of the issues involved. Importantly today, economic geography studies the consequences for people, countries and environments of economic activities.

The economic system

Geographers study the economy because it is the foundation of a country's wellbeing. A study of the economy is a study of what, where and for whom goods are produced. It is the study of the changing impact of activities on people and environments. As economies and nations have become more complex and inter-dependent, it is a study of the interrelationships that permit or hinder the social and economic development of countries.

The economic system is divided into four, and sometimes more, sectors:

- **primary** — that part of the economy concerned with the collection and use of natural resources
- **secondary** — the manufacturing or industrial sector — the part that processes resources into goods that people want
- **tertiary** — the sector that enables goods to be traded — it includes wholesaling, retailing, transport and entertainment, including tourism and personal services
- **quaternary** — the skilled **service sector**, including government, research and development, administration, banking, insurance and financial management — it involves the assembly, transmission and processing of information and knowledge

The balance of employment and contributions to a country's economy change over time. The modified Clark–Fisher model (Figure 3) attempts to show how this balance changes as a country develops.

Figure 3
The modified Clark–Fisher model

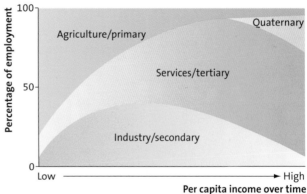

Types of economy

Geographers study why an activity is where it is — location — and how its locations have changed over time. They also study the changing interactions between the users of space in the form of transport, **trade**, flows of money and telecommunications. The interactions will be at different scales — from international deals supplying Australian coal to Japan, down to the purchase of a bottle of beer by an individual consumer. In studying economic geography, we are trying to explain *trends* in local, regional, national and international growth or decline, and *changes* in the location of economic activities. Changing location is the subject of Parts 5 and 6 of this book.

Geographers have also developed a critical stance. They examine the impact of economic change and development on peoples and environments. They assess whether activities are sustainable and whether the current world economy can continue to prosper in the future. The impacts of economic change are the focus of much of this book, including Parts 3 and 4.

The balance between the various sectors of the economy varies with the level of development of a country. Table 1 outlines the characteristics of the principle types of economy that you should be familiar with. The table does not show the increasing interdependence of the different economies, caused by trade, internationalisation and **globalisation**.

- **Trade** is the selling and buying of the resources for, and products of, the four sectors of the economy. As a general rule, the value added by goods and services rises from primary to quaternary. Trade is the subject of Part 7 of this book, with an in-depth case study of sugar.
- **Internationalisation** is the spread of economic processes across boundaries to create a more extensive pattern of economic activity.
- **Globalisation** involves the close economic interdependence between the leading nations in trade, investment and cooperative commercial relationships in which there are relatively few artificial restrictions on the cross-border movement of people, assets, goods or services. Globalisation and its impact is the focus of Parts 3 and 4. Part 8 looks into the future, in particular at what the nature of work will be like, and how sustainability can impact on economic activities.

The level of development underpins many explanations of economic systems (Table 1). Of particular current concern is the issue of 'making poverty history' (see p. 88), with the **G8** countries attempting to improve the lives of people in the poorest countries.

G8 leaders pose for a photograph with the President of the European Commission, José Manuel Barroso

EMPICS, Topfoto

Type of economy	Primary sector	Secondary sector	Tertiary sector	Quaternary sector
LDC/HIPC e.g. Burkina Faso, Rwanda, Tanzania	• Poor, mainly subsistence, agriculture • No forestry: deforested • Some fishing: poor • Declining energy resources • Few minerals	• Minimal manufacturing with some for export • World trade rules force industry to compete without capital • Patents prevent technology transfer, e.g. HIV/AIDS drugs	• Few services • Some growing tourism • Some government employment • Poor transport and communications	• Some government employment
LEDC e.g. Bolivia, Kenya	• Subsistence agriculture with some cash crops; plantation agriculture • Energy resources sometimes present; aid projects, e.g. dams, to be competitive	• Some processing of primary products and manufacturing for national needs • Industries forced by world trade rules	• Government service sector • Tourism • Poor transport and communications, except in some key areas • Outsourcing recipient	• Outsourcing of some financial activities
NIC/RIC e.g. South Korea, Brazil, Malaysia, China	• Agriculture and forestry: able to feed population • Able to acquire energy	• Large inward investment and locally generated investment • Industry protected as it establishes • Moving from work-bench to sophisticated production for export • Much TNC involvement	• Established tourism • Improving transport and communications, especially telecommunications	• Growing banking, insurance and investment activities, often state owned • IT strong • R&D developing fast
Former Soviet e.g. Hungary, Poland	• Once state controlled but now market oriented declining importance • Energy sector strong where resources exist	• Formerly state controlled and protected from competition • Now private or TNC controlled • Growing inward investment	• Capitalist services growing and former state institutions declining • Tourism growing	• Banking, insurance growing • Some R&D in stronger economies • In receipt of outsourcing from EU
Oil-rich OPEC e.g. UAE, Kuwait	• Abundant energy/ minerals • Small-scale agriculture, fishing and forestry	• Locally generated investment from oil wealth managed externally	• Some tourism in selected countries • Strong government involvement	• Strong banking, investment activities
MEDC e.g. Spain, Denmark	• Protected agriculture • Low contribution of forestry • Energy budget purchased or indigenous	• Strong manufacturing, which grew due to state protection • Declining basic industries • May hold patents	• Strong and dominant sector • Personal service growing • Tourism buoyant	• R&D strong • Knowledge-based economy • Strong IT
G8 (Canada, France, Germany, Italy, Japan, Russia, UK, USA)	• Protected agriculture and low contribution of fishing and forestry • High energy importer • Resource importer or resource rich	• Declining importance • Deindustrialisation • Inventive and innovative industries • Hold most patents and prevent technology transfer	• Very strong outward investment • Loss of jobs due to outshoring and outsourcing • Skilled labour shortages in personal services (e.g. doctors)	• Very strong banking, insurance and stock market orientation • Strong R&D • Strong administrative base of TNCs • Knowledge-based economy strong • Strong IT

Table 1 *The characteristics of the world's major types of economy*

At the lowest extreme of the levels of development (Figure 4) are the World Bank's 'low-income countries'. They are often referred to as **heavily indebted poor countries (HIPCs)**. More recently, the World Bank has focused on 30 **low income countries under stress (LICUS)**. Neither of these World Bank groupings is the same as the 'least developed economies' identified by the United Nations. Similarly, the World Bank's 'high-income countries' are more widespread than the G8.

❶ Haiti
❷ Cape Verde
❸ The Gambia
❹ Guinea-Bissau
❺ Guinea
❻ Sierra Leone
❼ Liberia
❽ Togo
❾ Benin

❿ São Tomé & Principe
⓫ Equatorial Guinea
⓬ Senegal
⓭ Mauritania
⓮ Burkina Faso
⓯ Mali
⓰ Niger
⓱ Chad
⓲ Central African Republic

⓳ Democratic Republic of the Congo
⓴ Angola
㉑ Zambia
㉒ Tanzania
㉓ Mozambique
㉔ Sudan
㉕ Ethiopia
㉖ Somalia

㉗ Djibouti
㉘ Eritrea
㉙ Madagascar
㉚ Lesotho
㉛ Malawi
㉜ Comoros
㉝ Burundi
㉞ Rwanda
㉟ Uganda

㊱ Yemen
㊲ Maldives
㊳ Afghanistan
㊴ Bhutan
㊵ Nepal
㊶ Myanmar
㊷ East Timor
㊸ Laos
㊹ Cambodia

㊺ Solomon Islands
㊻ Bangladesh
㊼ Tuvalu
㊽ Kiribati
㊾ Vanuatu
㊿ Samoa

Figure 4 *The world's 50 least developed countries, according to UNCTAD (UN Conference on Trade and Development), 2003*

The UK Department for International Development (DFID) has a list of **fragile states** — those countries where the government cannot, or will not, deliver core aspects of life to the population. These countries contain 14% of the world's population and their populations are growing at more than 1.5% per annum. Their incomes are stagnant, so they contain concentrations of the world's poorest people.

Three sectors of employment

The International Labour Office (ILO) has three broad groupings of employment:
- agriculture — **primary sector**
- industry — **secondary sector**
- services — **tertiary sector**

Of the 200 countries recorded by the ILO, only eight have industry as the dominant sector of employment. They are Albania, Bahrain, Bosnia-Herzegovina, Brunei Darussalam, the former Yugoslavian Republic of Macedonia, Niger, Northern Mariana Islands, and West Bank and Gaza.

Most countries tend to polarise between those dominated by agriculture — mainly in Africa — and those dominated by services. Asia and the Pacific have the most balanced employment — typically 40% agriculture, 28% manufacturing and 32% services. The economies with the largest share of employment in services are also those with the highest **GDP**. However, if the informal sector, such as barter and street trading, is included, there are many countries in the developing world where over 50% of the population work in that sector — for example, Kenya, Ghana, Paraguay, Peru and Pakistan.

1

Using case studies

Question

Study Figure 5. Use the information given in the text above to describe and suggest reasons for the relationships shown on the graph.

Guidance

Figure 5 is annotated to show you a planning sequence for describing graphs. A glossary of key terms is provided on pp. vi–x, because using the language of business and economics effectively is an important part of achieving success in your studies of geography.

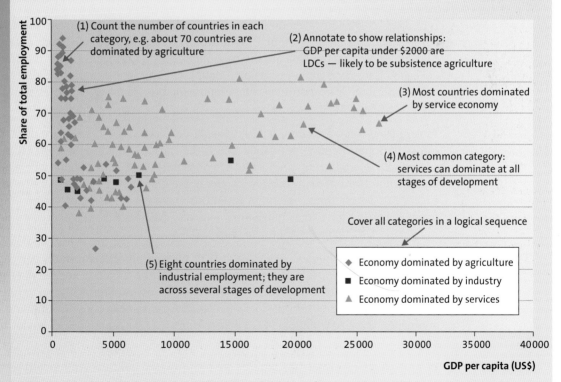

Figure 5 *Relationship between share of employment by sector and GDP per capita for various countries*

Part 2

An unequal world

- How have the patterns of wealth and poverty shown in Table 1 on page 3 come about?
- How can we measure and portray the variations in economic wellbeing and wealth?
- How can we explain the consequences of global variations in economic wellbeing?

Indicators of economic development

Indicators of global development are normally based on data from the **UNDP (United Nations Development Programme)** and the World Bank. Both organisations point out that the data are the best obtainable, despite flaws such as different collection dates and differing national definitions of the indicators. Table 2 shows some indicators of economic development using the groups of countries shown in Table 1 on page 3.

Table 2 shows that a country's rank varies if different single criteria are used. **GNI (gross national income)** per capita in $ is the most common indicator. GNI is the sum of all value added by production of goods and services in a country plus taxes and income from abroad, divided by the population. GNI identifies all of the G8 except Russia and draws attention to **MEDCs** such as Denmark, whose per capita figure is high but for a small population. Similarly, **NICs** such as Singapore and Hong Kong have **GNI** figures that equate with MEDCs.

A second, frequently used, measure is **GDP (gross domestic product)** per capita in $. GDP is the value added by all resident production of goods and services plus taxes within a country but excluding overseas earnings and profits that go to foreign investors. The ranking mirrors that for GNI and the same comments apply to the anomalies. Over 70% of global GDP comes from 11 MEDCs and former **Soviet bloc** countries.

PC and telephone ownership patterns are technological indicators of modern service economies and domestic affluence that reinforce the notion of a global divide.

Country	Category	GNI per capita (2003)	GDP per capita (2004)	PCs per 1000 people (2002)	Telephone/mobile phones per 1000 people (2002)	<$1 per day (%)	Employment in agriculture (%) (2002)		Employment in industry % (2002)		Employment in services (%) (2003)		% of females in labour force (2002)
							M	F	M	F	M	F	
USA	G8	37 610	37 800	659	1 134	0	3	1	32	12	65	87	46.2
Japan	G8	34 510	28 200	382	1 195	0	5	5	37	21	57	73	41.7
UK	G8	28 350	27 700	406	1 431	0	2	1	36	11	62	88	44.3
Germany	G8	25 250	27 600	431	1 378	0	3	2	44	28	52	80	42.4
Canada	G8	23 930	29 800	487	1 013	0	4	2	33	11	64	87	46.0
France	G8	24 770	27 600	347	1 216	0	2	1	34	13	64	86	45.3
Russia	G8	2 610	8 900	89	362	6.1	ND	ND	ND	ND	ND	ND	49.2
Denmark	MEDC	33 750	31 100	577	1 522	0	5	2	36	14	59	85	46.5
Spain	MEDC	16 990	22 000	196	1 330	0	8	5	42	15	51	81	37.5
Kuwait	Oil-rich	16 340	19 000	121	723	0	2	0	36	3	62	97	32.1
UAE	Oil-rich	9 080	23 200	129	1 010	0	9	0	36	14	55	86	15.9
Hungary	Former Soviet	6 330	13 900	108	1 037	<2	9	4	42	26	49	71	44.8
South Korea	NIC	12 020	17 800	556	1 168	<2	9	12	34	19	57	70	41.8
Malaysia	NIC	3 780	9 000	147	567	<2	21	14	34	29	45	57	38.3
Brazil	NIC	2 710	7 600	41	424	8.2	24	16	27	10	49	74	35.5
China	NIC	1 100	5 000	46	328	16.9	ND	ND	ND	ND	ND	ND	45.2
Kenya*	LEDC	390	1 000	6	52	23.0	20	16	24	10	57	75	46.1
Burkina Faso	LDC/HIPC	300	1 100	2	13	44.9	92	93	3	2	5	5	46.5
Tanzania	LDC/HIPC	290	600	4	24	19.9	ND	ND	ND	ND	ND	ND	49.0
Rwanda	LDC/HIPC	220	1 300	ND	16	35.7	88	98	5	1	7	1	48.7

Note that the employment data do not add up to 100% because some sectors are not included.
Employment is % of male or female employed population. ND = no data.

The World Bank data on the proportion of the population earning below $1 a day are stark. One of the eight Millennium Development Goals is to reduce extreme poverty. Only **LEDCs/HIPCs** and the former Soviet bloc have significant percentages earning so little. The worst cases, not shown in Table 2, are Mali (72.8%), Zambia (63.8%) and Nigeria (61.4%).

Table 2 Indicators of economic development

Brandt's 'north–south' divide

In 1980 the *Brandt Report* drew attention to the gross economic inequalities in the world, which were characterised by the simple concept of the rich 'north' and the

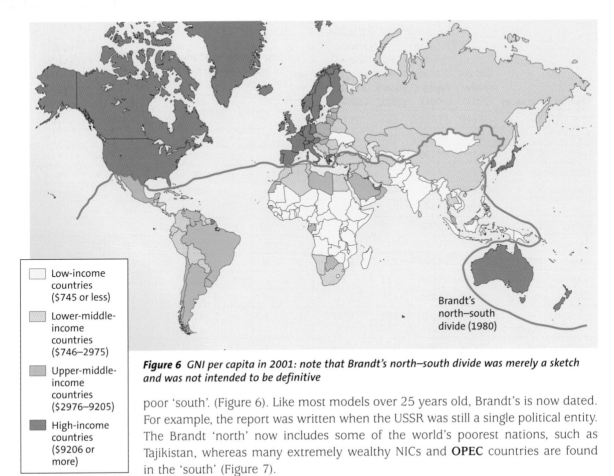

Low-income
countries
($745 or less)

Lower-middle-
income
countries
($746–2975)

Upper-middle-
income
countries
($2976–9205)

High-income
countries
($9206 or
more)

Brandt's
north–south
divide (1980)

Figure 6 *GNI per capita in 2001: note that Brandt's north–south divide was merely a sketch and was not intended to be definitive*

poor 'south'. (Figure 6). Like most models over 25 years old, Brandt's is now dated. For example, the report was written when the USSR was still a single political entity. The Brandt 'north' now includes some of the world's poorest nations, such as Tajikistan, whereas many extremely wealthy NICs and **OPEC** countries are found in the 'south' (Figure 7).

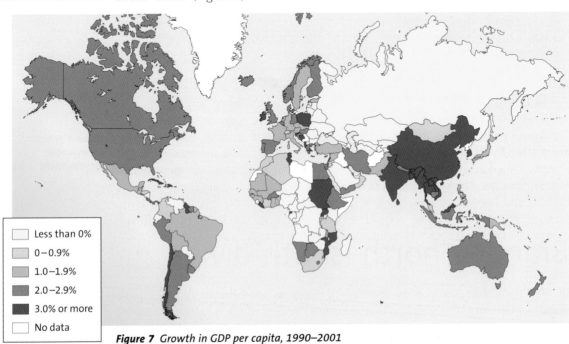

Less than 0%

0−0.9%

1.0−1.9%

2.0−2.9%

3.0% or more

No data

Figure 7 *Growth in GDP per capita, 1990–2001*

Question

Use Figures 6 and 7, together with Table 2, to redraw the Brandt line to reflect the current position.

Guidance

Start with the anomalies and then consider Figures 6 and 7. Which countries appear on the 'wrong' side of the 1980 line? Should there be outliers, such as UAE?

The *Brandt Report* made recommendations to improve matters, but none of these has ever been fully implemented. The levels of debt of many countries have increased since Brandt proposed measures to reduce debt.

Increasing debt burdens

Debt among the least developed countries has increased. Why is this so?

- When money was originally lent to developing countries, export revenue from raw materials and crops was more than adequate for the repayment of interest. However, currency fluctuations mean that LEDCs with far less currency are more vulnerable to rapid reductions in the value of a currency.
- The price of many commodities has fallen over the years, making it harder for LEDCs and **LDCs** to repay their loans. Price fluctuations are hard for LDCs to manage.
- Many countries struggled to repay mounting interest out of diminishing export revenue and have been obliged to take out further loans. These new loans were tied to **structural adjustment programmes (SAPs)**, which imposed strict spending conditions on the receiving countries. These conditions obliged the receivers to cut back on healthcare, education, sanitation and housing programmes, causing a further deterioration in the quality of life of the population.
- The *Brandt Report* recommended convening a summit of world leaders to plan a major international relief programme, targeting hunger and poverty. The countries most affected by infectious diseases lack the money to implement disease control programmes because of the cutbacks.
- The *Brandt Report* recommended that there should be a commitment to raise the income and quality of life of people in developing nations, rather than allowing companies to invest and produce mainly where wages, taxes, trade restrictions, financial regulations and environmental safeguards are low. Companies are allowed to profit from cheap labour in the developing world, but are not obliged to ensure the health and wellbeing of the workers. These double standards were employed by the Victorian industrialists in Britain in the late 1800s. Today, such double standards are shameful.
- The *Brandt Report* stressed the importance of preserving the environment. The developed world does far more damage to the environment but the less developed world suffers more and has less money to counteract environmental destruction, often caused by TNCs.

The consequence of increased debt is a complex pattern of wealth and poverty, best illustrated by GNI per capita (Figure 6). Figure 7 shows that the economies of the world are growing at differential rates. The highest growth is generally among countries in south and east Asia where **NICs (newly industrialising countries)**, for example Malaysia, and **RICs (recently industrialising countries)**, for example Cambodia, were relatively unknown in 1980. The countries with the lowest 'growth', which can be seen to be negative (i.e. a reduction), are found in the African continent and in much of the former **Soviet bloc**. Changes in the former Soviet bloc are a consequence of post-1989 developments. Even so, the economic wealth of the former Soviet bloc is far higher than other areas of the world, especially in terms of minerals and energy resources.

The development gap

Figure 7 does not take account of the wealth that already existed in 1990 and, therefore, can be misleading. For example, Switzerland is shown as having slow growth but cannot be considered to be a poor country. The gap between the richest countries (the 'haves') and the poorest (the 'have nots') is known as the **development gap**. The development gap is not a spatial divide, like the north–south divide. There is disagreement over what is actually happening to the development gap; is it widening or is it diminishing?

Is the development gap narrowing or widening?

At a global scale, the 'haves' are the richest 20% of countries, which consume about 80% of all resources. These countries have stable political systems that people vote for, freedom of speech, human rights and the right to education and healthcare. The 'have nots' are the poorest 20% of people, who earn only 1.3% of global income (Figure 8) and lack some, or all, of the other benefits listed above.

Note, however, that wealthy countries do not wholly consist of 'haves' and that even the poorest countries have their share of wealthy people. Because of these disparities, most countries have *internal* development gaps. Increasingly, development is seen as an all-round concept, including not only economic growth but also social development.

Figure 8 The human development index for five countries and the share of national income of the richest and poorest

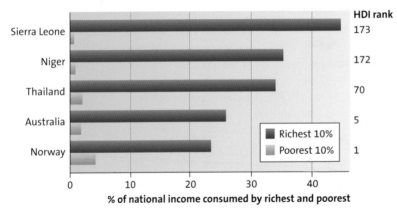

Figure 8 shows the **human development index (HDI)**, which takes a broader view of development, for five countries. LEDCs have the worst income distribution, with 10% of their population surviving on 1–2% of income, while the 10% wealthy elite get over 40% of national income. The picture is more balanced in **MEDCs**, but still unequal. In many **LEDCs**, the development gap is an urban versus rural split because urban areas tend to have better services and more opportunities.

Who bridges the gap?

The news about the development gap is not all bad. Some countries are in the process of successfully 'bridging the gap'. Singapore has developed so far that it should now be classified as an MEDC. Other NICs have undergone staggering economic growth in the last few decades, which has seen millions of people move from poverty to relative prosperity. The UN describes the progress made in east Asia since the 1970s as 'one of the largest decreases in mass poverty in human history'. As business globalised and TNCs moved into east Asia, jobs and relative prosperity followed. This has not been without its problems — potential exploitation of workers, low wages and market crashes (such as 1998) are all legitimate criticisms (see Part 4).

Economic growth is the main way in which the development gap will be narrowed. Unfortunately, growth is not evenly distributed. Figure 9 shows the powerful G8 countries continuing to see their wealth grow, while NICs have seen the most rapid growth as they benefit from globalisation and outsourcing. Some RICs, such as Thailand, and some LEDCs are growing more slowly. The African LDCs are hardly growing at all and income growth in many former communist countries has been negative. Much of the world's wealth is in the hands of the G8/MEDC nations and most TNCs originate in them. MEDCs also have the controlling stake in institutions such as the IMF (International Monetary Fund) and the World Bank, and 80% of world trade occurs between these countries.

Figure 9 World income: growing inequality, 1820–2010 (projected)

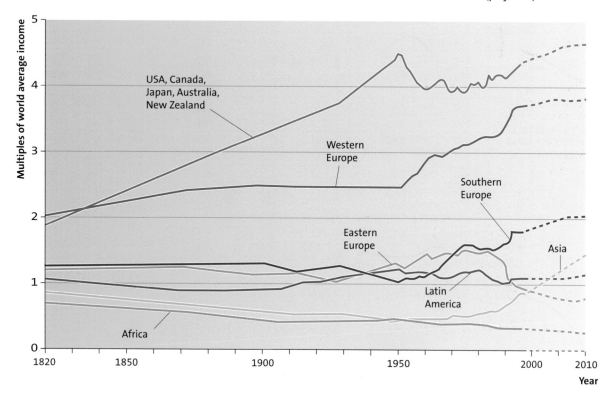

Question

Use Figure 9 (which shows trends by the world's regions) and Table 3 to discuss whether the development gap is widening or narrowing.

Guidance

- Annotate the graph in Figure 9 to summarise the key trends. For example, there are clearly groups of countries that are improving relative to other groups.
- Use two colours to highlight increases and decreases in development indicators in Table 3. Note that an increase in infant mortality is *not* an improvement.
- Use this evidence to support your arguments for widening and narrowing of the development gap. The question is complex and much depends on what indicators are used and which countries are chosen.

Country	Wealth		Population		Health				Education			
	GNP per capita ($US)		Life expectancy at birth (years)		Infant mortality (per 1000 live births)		Adult literacy (%)		Primary school enrolment (% male/ female)		Energy consumption per capita (kg oil equivalent)	
	1989	1999	1989	1999	1989	1999	1989	1999	1989	1999	1989	1999
USA	21790	29240	76.0	76.8	11	7	99.9	99.9	96/94	94/95	7882	8040
Jamaica	1500	1740	73.2	75.1	19	10	85	86.4	85/83	89/87	931	1289
Brazil	2680	4630	66.0	67.5	68	34	80	84.9	82/80	100/96	915	671
Sierra Leone	240	140	42.2	38.3	202	182	35	32.0	51/40	59/41	77	560
Ghana	390	390	55.1	56.6	133	63	54	54.6	77/72	82/72	68	30
Kenya	370	350	58.2	51.3	106	76	77	81.5	77/76	92/89	100	89
UK	16100	21410	74.5	77.5	9	6	99.9	99.9	98/98	97/98	3646	3938
Russia	2620	2260	67.2	66.1	14	18	99	99.5	95/95	93/93	4320	4230
Saudi Arabia	7000	6910	64.5	71.3	79	20	74	76.4	93/92	91/93	5033	4528
India	350	440	59.0	62.9	118	70	52	56.5	55/49	78/64	231	297
Bangladesh	210	350	50.1	58.9	155	58	37	40.8	40/37	80/83	57	76
Thailand	1420	2160	65.2	69.9	35	26	89	95.3	72/71	82/79	352	1036
Vietnam	260	350	66.2	67.8	51	31	81	93.1	70/68	95/94	100	149
China	370	750	70.0	70.2	31	33	81	83.5	91/01	99/99	598	709
World	4890	4890	66.1	66.7	66	56	–	–	–	–	–	–

Table 3 Some key development indicators compared, 1989 and 1999

Development indicators are not solely economic, as Table 3 shows. Today, social scientists include other aspects of human development, such as literacy, education levels, health indicators and quality of life measures.

Globalisation and global shift

Globalisation is a common term in modern business and economic geography. The 2004 Asian tsunami disaster served to stress **global interdependence** — the way in which economies and societies are interlinked. A problem in one country can have repercussions in others. For example, the tsunami destroyed hotels in Phuket, many of which depended on European tourists and were owned by Western companies.

> Globalisation is the generic term for the process of integration in the realms of trade, economic relations and finance (it is broader, including social relations, knowledge, culture and politics) and it is not new. It has been aided by the ICT revolution that has destroyed distance and indeed time. Brands are known the world over and are potentially destroying local diversity.
>
> Source: *Financial Times*

Globalisation and the evolution of capitalism

Colonialism and **mercantilism** were 'global' processes financing the Industrial Revolution. The Industrial Revolution led to market economies backed by power and dependent on 'empires' — for example, the British, French and German empires.

Between 1918 and 1945, the Communist Revolution, the Depression, and the rise of **Keynesian interventionism** in the 1930s led to a new world economic order. The World Bank, GATT (the General Agreement on Tariffs and Trade) and the Bretton Woods Agreement (1948) were the foundations of the new system after 1945. Trade was governed by conventions that supported the economies of the wealthy nations.

The origin of modern globalisation can be found in the 1975 OPEC oil price rises. The new wealth of the oil producers was invested in MEDC banks which, in turn, loaned the money to both the developing world and emerging industrial economies with cheaper labour costs. The fall of the Soviet bloc after 1989 led to further opportunities for investment. State-owned enterprises were privatised and often sold to outside interests. Global corporations have become more important than governments, controlling greater wealth and investment thanks to support from the **World Trade Organization (WTO)**.

Globalisation takes three forms:

- economic — growth of TNCs at the expense of national governments
- cultural — increased Western influence, especially American, over aspects of culture, for example music (downloaded from the internet to iPods), the arts (films), and the media (e.g. *CNN* and *Sky News*)
- political — the increased influence of Western democracies and the decline of centralised Soviet economies

Geographers like to add a fourth form: environmental globalisation (see Figure 10).

Factors promoting globalisation

Globalisation differs from **internationalisation** (which is defined as 'the extension of economic processes across national boundaries'). Globalisation involves not just the geographic extension of economic activity across national boundaries but, primarily, it involves the integration of internationally dispersed activities within single companies. Some people see globalisation as a cycle that returns some economic power to those countries that lost it as a result of colonialism. Others see it as a set of linear stages in the process of economic development and domination of the world economy by the wealthy.

The following points outline the factors behind the process of globalisation:

- money from the OPEC 1975 price rise (see above)
- computer technologies (ICT) that enabled (1) rapid movement of money and information and (2) development of computerised production and robotics, replacing skilled labour and making production more mobile
- fibre optic cables and satellites that could transmit more information
- the development of the internet and the world wide web
- cheaper and quicker transport (e.g. large bulk carriers and jumbo jets)
- the strength of the WTO and **trade blocs** promoting free trade
- the rise of **multinational companies (MNCs)** responsible for 80% of foreign investment and 50% of world trade
- new products (e.g. G3 mobile phones) and new production methods
- increased importance of services such as insurance operating at a global scale

Views about globalisation

Many economists argue that globalisation has widened the development gap (see p. 10). A 1999 report attributed the debt crisis to the effects of globalisation on the economies of the poorest states, whereas MEDCs had become richer. Companies relocate production and services to countries where labour is cheap while retaining control and R&D in the home country. The process of globalisation is part of a raft of economic policies that use the rhetoric of open markets and free trade to mask a growing gap between rich and poor, both within and between countries.

Environmentalists, human rights advocates, trade unionists, dispossessed farmers from developing countries and citizens' groups decry globalisation and sometimes protest violently at summits of the world's most powerful leaders — for example, in Seattle (2000), Cancun (2003) and Davos (2004). At the same time, economists, business leaders, TNCs and global institutions (such as the World Bank), praise globalisation as a historical inevitability. Figure 10 summarises some of these conflicting views and shows how divided global opinion is.

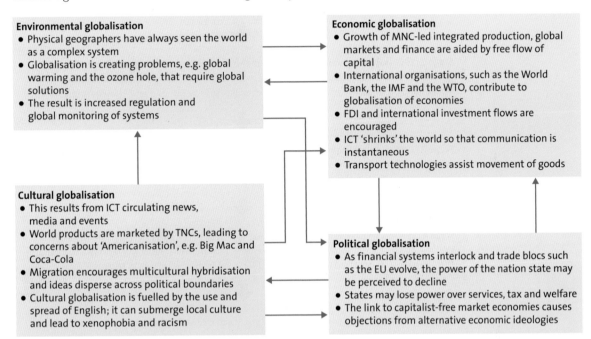

Environmental globalisation
- Physical geographers have always seen the world as a complex system
- Globalisation is creating problems, e.g. global warming and the ozone hole, that require global solutions
- The result is increased regulation and global monitoring of systems

Economic globalisation
- Growth of MNC-led integrated production, global markets and finance are aided by free flow of capital
- International organisations, such as the World Bank, the IMF and the WTO, contribute to globalisation of economies
- FDI and international investment flows are encouraged
- ICT 'shrinks' the world so that communication is instantaneous
- Transport technologies assist movement of goods

Cultural globalisation
- This results from ICT circulating news, media and events
- World products are marketed by TNCs, leading to concerns about 'Americanisation', e.g. Big Mac and Coca-Cola
- Migration encourages multicultural hybridisation and ideas disperse across political boundaries
- Cultural globalisation is fuelled by the use and spread of English; it can submerge local culture and lead to xenophobia and racism

Political globalisation
- As financial systems interlock and trade blocs such as the EU evolve, the power of the nation state may be perceived to decline
- States may lose power over services, tax and welfare
- The link to capitalist-free market economies causes objections from alternative economic ideologies

Figure 10 The web of globalisation and its possible outcomes

Global shift is one consequence of the latter stages of globalisation (Figure 11). Global shift involves the movement of economic activity from MEDCs, originally to NICs, and then to RICs and LEDCs. Initially, the shift involved labour-intensive manufacturing, but increasingly it has involved all sorts of manufacturing and, more recently, services.

Outcomes of globalisation and global shift include the following:

- **Less regulation of companies** — so much so that private companies may have the same rights as national governments in countries receiving inward investment.
- **Privatisation** — this is attractive to globalising companies as it leads to lower labour costs. There is a danger of demand declining in MEDCs because wages are kept down by the new, cheaper global locations. For example, wage rises in Germany have slowed to enable firms to compete with the new producers in eastern Europe.
- **Deregulation of financial markets** — so that currencies become more volatile, to the detriment of many LEDCs.
- **Falling returns on primary products** — reducing the income of countries supplying raw materials and food.

Figure 11
The stages of global shift

Stage 1 (1945–1960) Activities that dominated between 1875 and 1940 (e.g. textiles) declined.   New activities went to new sites (e.g. in New Towns) and old industrial buildings were abandoned — unsuitable for modern technologies. Governments ameliorated the effects of decline with regional development aid, but this rarely replaced the jobs or retained traditional skills. The cycle of production rising, peaking and declining as new products are invented (the **product life cycle**), goes along with changes in location and employment characteristics. Changing production cycles are a component of **deindustrialisation** — the decline of regionally important industries, as in Saarland (see Part 4) — with **reindustrialisation**.

Stage 2 (1960s) Heavy industry developed in the Far East (e.g. Japanese and South Korean governments backed shipbuilding and steel production). This led to early deindustrialisation. Textiles and other labour-intensive industries began to move to Asia and elsewhere (e.g. Turkey).

Stage 3 (1980s) Automobile assembly shifted, which started with inter-war internationalisation (e.g. Ford in Europe). Japanese car makers established in the EU and elsewhere. The early stages were shifts within the MEDCs; NICs became involved at a later stage.

Stage 4 (1980s, 1990s) Electronics branch plants shifted to NICs and RICs (e.g. Malaysia and Singapore). There were some inter-NIC moves (e.g. Taiwan to Malaysia and Malaysia to Vietnam). The ultimate shift occurred when NIC-based TNCs set up branch plants in MEDCs.

Stage 5 (Post-1989) This era saw the rise of privatised industrial production in the former Soviet bloc. Many eastern European regions became sites of regeneration and the global shift (see *Case study 2* on pp. 22–26).

Stage 6 (1990s) Financial services and banking moved to NICs as part of the global service sector working around the clock. Outsourcing began, largely to India.

Stage 7 The rise of service industry outsourcing or business process offshoring (BPO) to include call centres, back office functions and other routine business support operations (see Part 6).

Stage 8 Relaxation of trade rules (e.g. for textiles) further reduces MEDC output. The search for ever-cheaper locations leads to shifts from NICs to RICs and LEDCs for a range of goods (e.g. Chinese shoes) and services.

Stage 9 The growth of inward investment to China, aided by state control of foreign investment and a fixed currency exchange rate. Focus is primarily industrial.

- **'The global casino'** — international finance, including share and currency traders, and foreign exchange (Forex) and derivatives markets have increased wealth.
- **Environmental consequences** — such as deforestation and pollution.
- **A widening development gap** — the main economic consequence, both internationally and within nations. Western companies buy into the tiger economies and, more recently, India and China (see Parts 4 and 8).
- **Cultural consequences** — for example, economic migration, refugees and asylum seekers — as peoples search for the economic wellbeing that they view enviously from their LEDCs.
- **Geopolitical consequences** — such as moves towards greater unity of economies — for example, the establishment of the Euro zone and trade blocs. At the same time, protest movements (Seattle and Genoa), revolution (Indonesia) and terrorism (9/11) are seen as opposition to the political manifestations of economic globalisation.

The role of large companies in globalisation and global shift

Large companies can be divided into three groups (see Table 4).

	Transnational companies (TNCs), sometimes called MNEs (multinational enterprises)	Multinational companies (MNCs)	Global companies
Where they operate	More than one country	Number of separate, disparate units, on a semi-independent basis	In different countries, but all operating to a single plan
How they operate	Separate divisions in each country, with either one HQ or sometimes dual HQs (e.g. Shell)	Units operate independently for purchasing and sales	Manage purchases centrally to obtain economies of scale Fewer production units (e.g. General Motors car works closed in Luton in 2002) 'Badge engineering' (same product with different badges to suit local markets)
Size	Smaller than an MNC	Single product group	Large scale
Organisation	Less complex	Operates through subsidiaries	Centrally managed multiproduct
Example	Shell, before it became global	General Motors was a MNC in 80 countries, e.g. Vauxhall car plant in Luton, UK, with HQ in Detroit	General Motors became a global company (GM) in the 1990s

Table 4 Types of large company

In 2005, Procter and Gamble merged with Gillette to form a new **global company** (P&G), producing a multitude of household products. Figure 12 illustrates the diversity of this global company and the worldwide nature of its sales.

The *Financial Times' Annual Global 500 Report* ranks the world's largest companies by market value. (A ranking by numbers employed would produce a different list.) The start of the 2005 listing is shown in Table 5.

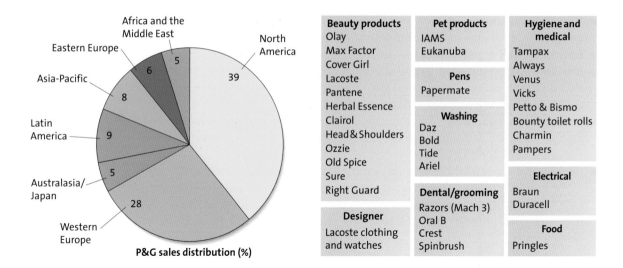

The dominance of American companies in Table 5 is apparent. The major global companies are increasingly service oriented, focusing on IT, health, banking, retailing and beverages. Only General Electric (the largest company) and Toyota (in position 19) are 'industrial' in the traditional sense. The HQ location for selected types of industrial and service industries demonstrates the importance of American-based companies (Table 6).

Table 5 The top 20 companies in the Global 500 Report, 2005

Rank	Company	Country of HQ	Product/sector
1	General Electric	USA	Diversified industrial
2	Exxon Mobile	USA	Oil and gas
3	Microsoft	USA	Software and computer services
4	Citigroup	USA	Banking
5	BP	UK	Oil and gas
6	Wal-Mart Stores	USA	Retailing
7	Royal Dutch/Shell	Netherlands/UK	Oil and gas
8	Johnson & Johnson	USA	Pharmaceuticals and biotechnology
9	Pfizer	USA	Pharmaceuticals and biotechnology
10	Bank of America	USA	Banking
11	HSBC	UK	Banking
12	Cisco Systems	USA	Information technology hardware
13	Vodafone Group	UK	Telecommunications services
14	IBM	USA	Software and computer services
15	Total	France	Oil and gas
16	Intel	USA	Information technology hardware
17	American International Group	USA	Insurance
18	Altria	USA	Tobacco
19	Toyota	Japan	Automobiles and parts
20	GlaxoSmithKline	UK	Pharmaceuticals and biotechnology

Source: *Financial Times*

Grouping	USA	UK	Japan	Germany	France	Other EU + Norway and Switzerland	NIC/RIC
Automobiles	4		5	3	2		
Aerospace	7	1			1		
Chemicals	4		1	2	1		2
Steel and metals	1		2			1 Netherlands	1
Electronics	1		7	1		1 Netherlands	1
IT	19		7		2	1 Sweden 1 Finland	2
Software	8		4	2	1		
Telecoms	8	3		1	1	1 Estonia 2 Italy, 1 Netherlands 1 Portugal 1 Norway	7
Biotech	13	2	1		2	2 Switzerland 1 Denmark	
Beverages	3	2				1 Belgium 1 Netherlands	
Food	11	4	1		1	2 Netherlands 1 Switzerland	
Banking	16	7	4	2	2	3 Estonia 2 Ireland 3 Italy 1 Denmark 1 Spain	5
Insurance	13	3	2	2	2	2 Netherlands 2 Italy 2 Switzerland	

Table 6
HQ locations for selected industries and services

Using case studies

4 Question

Using a world map, indicate the HQ distribution and type of the major global companies listed in Table 5.

Guidance

A useful technique to use would be located proportional bars.

Tables 5 and 6 show how the new economic activities related to the provision of services for people dominate the global company listing. Only in the broad fields related to IT and telecommunications have NICs/RICs developed many global companies. Global banking companies in the NICs are confined to Singapore (2), Hong Kong (2) and South Korea (1). The position may be radically different by 2015, with the rise of China and India.

How TNCs operate globally

The following case studies show how companies benefit from globalised operations.

SOME GLOBAL COMPANIES

Intel

In 2005, Intel was the sixteenth largest company in the world (see Table 5). Intel has been a global company, making semiconductors and microprocessors to power desktop computers, for over 30 years. The global pattern of Intel's activities has always been market orientated and has shifted with the changing markets.

In 2004, Intel's product design took place in the USA and Asia; 75% of manufacturing took place in Oregon, New Mexico, Colorado and Massachusetts (USA) and 25% in Ireland and Israel. The company uses a 'copy exact' system of production, so every chip factory is set up in the same way. Assembly and testing take place in Costa Rica, Malaysia, China and the Philippines. Of the company's 100 000 workforce, 75% are in the USA, but this proportion is declining because the markets have shifted to Japan, southeast Asia and, more recently, China. As a result, Intel is expanding its output in the Far East, close to the biggest potential markets of China, Japan and India.

Despite being spread across the globe for marketing and cost-saving reasons, Intel still values face-to-face meetings and teleconferencing because it has yet to find a substitute for human interaction when problem solving. The cost of this policy is covered by the labour savings generated by Intel's global spread of production.

Figure 13
The global origins of the parts of a Dell computer

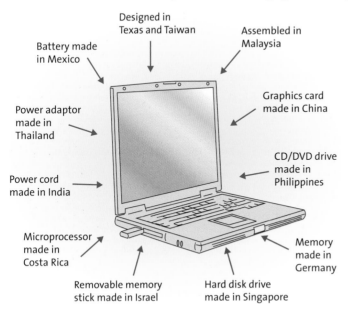

Battery made in Mexico

Power adaptor made in Thailand

Power cord made in India

Microprocessor made in Costa Rica

Removable memory stick made in Israel

Designed in Texas and Taiwan

Assembled in Malaysia

Graphics card made in China

CD/DVD drive made in Philippines

Memory made in Germany

Hard disk drive made in Singapore

Dell

The pattern of assembling a product from a range of parts manufactured across the world is common among computer manufacturers. Dell, like Intel, carries out its design in the USA but the subassemblies are made in NICs and RICs, where labour costs are cheaper (Figure 13). Sales in the UK are solely by mail order from a depot in Ireland. As a result, Dell is able to keep prices down and sales soaring. As a highly competitive company in the computer field, Dell reviews the location and organisation of its factories and assembly plants on an annual basis. Dell is at the forefront of subcontracting (especially design to Taiwanese companies).

Airbus

The Airbus A380, the largest passenger plane in the world, is assembled in Toulouse (France). There are a number of subassembly sites in Europe, but the company actually depends on a worldwide network of component suppliers in the developed world (Figure 14). Start-up costs were provided by the participating European governments.

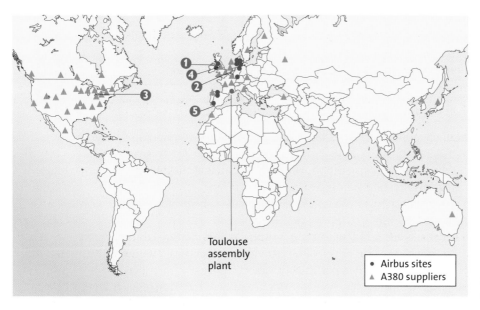

Legend for Figure 14:

❶ Broughton, Flintshire, UK — wings

❷ Hamburg, Germany — fuselage sections, tail fin

❸ Oakville, Ontario, Canada — landing gear

❹ Derby, UK — engines

❺ Puerto Real, Spain — tailplane

Toulouse assembly plant

• Airbus sites
▲ A380 suppliers

Figure 14 *Airbus A380 suppliers*

The globalisation of operations is different compared with Dell. Airbus uses subcontractors such as Rolls-Royce for specialist components. Rolls-Royce makes aero engines for all the large airlines of the world.

Global designer brands

Figure 15 shows the groups that make up the designer fashion industry. Many big names are parts of multinational groups that finance their activities. The big five control many names. The largest, Louis Vuitton Moet Hennessy (LVMH), produces other lifestyle products such as champagnes and brandies. Fashion is also managed by private equity groups, such as Texas Pacific, and industrial-owning groups, such as Marzotto. Here, the global interconnections are based on the need for worldwide sales.

Figure 15
The designer brands groups

Prada Group	Puig	PPR	LVMH		Compagnie Financière Richemont
Brands	**Brands**	**Brands**	**Brands**		**Brands**
Prada	Paco Rabanne	Gucci	Louis Vuitton	Fendi	Cartier (50% sales)
Miu Miu	Caroline Herrera	Yves Saint Laurent	Loewe	Stefano Bi	Chloé
Helmut Lang	Nina Ricci	Alexander McQueen	Celine	Emilio Pucci	Piaget
Jil Sander		Balenciaga	Berluti	Thomas Pink	Van Cleef & Arpels
Azzedine Alaia		Sergio Rossi	Kenzo	Donna Karan	Dunhill
		Bottega Veneta Boucheron	Givenchy	Dior	Hackett
		Beaute	Marc Jacobs		Lancel
		Bedat & Co			Old England

IT Holding	AEFFE	Texas Pacific Group	Others	Purdey
Gianfranco Ferré	Alberta Ferretti	Bally	Tommy Hilfiger	Shanghai Tang
Malo	Moschino		Karl Lagerfeld	
Exté		Hicks, Muse, Tate & Furst	Hermès	
	Fin.Part	Jimmy Choo	Jean Paul Gaultier	
Marzotto	Cerutti		Diesel	
Hugo Boss			Martin Margiela	
Marlboro Classics	Phillips-Van Heusen	Falic Group	Wang Xiao Lan	
Valentino	Calvin Klein	Christian Lacroix	Lanvin	

▨ Major holding groups
▨ Industrial holding companies
▨ Financial group holdings

Question

With reference to a range of examples:
(a) discuss the factors that are assisting the process of globalisation
(b) examine the advantages and disadvantages of globalisation

Guidance

This is a good example of how questions on economic geography can overlap several sections of any textbook.

- Define 'globalisation' in your introduction. Keep to 'economic globalisation' as the main focus. The question is open, so define your focus.
- Define terms (e.g. global shift) as you go along. It convinces the examiner that you know what you are talking about.
- This question is not just about manufacturing, so you can include services (see Part 6).
- The 'range of examples' could be of economic activities in countries. The rise of NICs and RICs (see Part 7) could be illustrated by a particular country.
- Structure your answer to include the role of TNCs. Use any of the companies from this case study. Note the role of colonialism, free trade, improved transport and communications.

Case study 2 — ## VOLKSWAGEN: A GLOBAL MANUFACTURING CORPORATION

The Volkswagen Audi Group (VAG) is the world's fourth largest motor vehicle production company. In 2002, it produced 12% of the world's cars. The company currently has 44 assembly plants in 11 European countries and in seven countries in the rest of the world. In 2006, the group employed 300 000 people, of whom 103 000 worked in Germany. On average, 14 000 vehicles are produced every working day. Since the Second World War, VAG, the parent company, has become a truly global **transnational company**.

The original location of VW in Wolfsburg was a strategic decision as Germany built up to the Second World War, because it was at the known limit of bomber range from both west and east. Despite the involvement of the company in wartime production, it became a key element in the post-1945 recovery of the West German economy. VW was nationalised in 1949 and owned by the national and regional governments. To accomplish its post-war role, VW had to export to earn funds to aid reconstruction and to develop its markets to earn money to reinvest in expansion. The VW Beetle became a symbol of West German economic recovery. The company's labour record, being strike-free, helped VW maintain its production and satisfy demand.

There are several key factors explaining the development and spread of VW. The company used seven strategies as it expanded to become a European-based global manufacturer (see Figure 16).

Strategy 1: to grow in West Germany

Early growth was located entirely in Nieder Sachsen, the federal state that owned much of the company. A

Figure 16 *Summary of VW's strategies for expansion*

1 and 2 Take over Audi in West Germany

3 Take over SEAT in Europe

4 Global expansion

VW GROUP EXPANSION STRATEGY

7 Effective use of globalisation for technological innovations

5 Expand into eastern Europe

6 Develop luxury cars

component factory in Braunschweig and plants in Hanover and Kassel were all relatively close to the main assembly plant at Wolfsburg. Emphasis was placed on rail communication between the plants, which has been maintained ever since as the logistics basis of the group's European locations. The construction of an assembly plant at Emden followed the trend common to most European automobile manufacturers in the 1960s of opening in development areas where government grants aided investment. The Salzgitter plant was located to bring alternative employment to an area suffering from declining employment in iron mining and steel.

Strategy 2: acquiring other companies in West Germany

The acquisition of Auto Union and NSU and the subsequent creation of the Audi brand in 1965 was a major move designed to acquire manufacturing capability in the south of West Germany. Subsequently, Audi became a separate brand group within VAG, focusing on higher-quality, more expensive cars rather than the quality mass market output of VW.

Strategy 3: acquiring major producers in Europe

In order that VW could compete further in Europe, the next strategy was to acquire other car makers. As with many activities, these takeovers were often started in collaboration with other companies. In 1982, a cooperation agreement was signed with the Spanish car maker SEAT, which led to the takeover of the company in 1985 and its subsequent embedding in the Audi brand group. Cooperation with Scania in Sweden gave the group a foothold in heavy vehicle manufacture.

Strategy 4: expansion on each continent

Because VW achieved such success in global markets in the post-war period, it was logical for the company to consider manufacturing in its markets rather than transporting finished vehicles across the globe. It was a strategy that gave access to markets where tariff barriers made products manufactured in West Germany expensive. For this reason, the company opened a plant at Puebla, Mexico, in 1967. The opening of assembly plants in South Africa and Brazil provided access to markets on two more continents. The components initially came from West Germany. More recently, these plants became part of a global components supply system. Production in Brazil, in particular, has expanded by acquisition of a share in developments with other auto makers, and partly by developing new sites, such as Curtiba.

In the 1980s, VW took an advantage over other car makers when it began to invest in China. Twenty years later, this investment has led to the VW group being one of the largest car suppliers in China. In 2000, the group opened an assembly plant for Skoda in India, thus beginning the process of supplying another untapped market.

Strategy 5: expansion into eastern Europe, mainly after 1989

Of all the global car makers, VW has invested most rapidly in the former Soviet satellite countries. The state-owned enterprises in eastern Europe were in need of investment to enable them to compete in the world marketplace and to enable them to develop modern products. The takeover of the Czech/Slovak company Skoda in 1990, together with the purchase of the former East German Trabant plants at Chemnitz and Mosel, enabled the company to shift some production to cheaper labour locations and to re-launch the Skoda brand as a mass market producer. The strategy of using locations with lower labour costs was repeated with the opening of Audi's plant at Gyor in Hungary and the opening of component-making plants in Poland.

Topfoto

Strategy 6: developing an up-market arm making luxury cars

The rebranding of Audi as a luxury car maker was a further strategy to broaden the market for the group's products. In 1998, the group bought Rolls-Royce/Bentley — the British luxury car maker. VAG had the rights to manufacture Rolls-Royce but not the right to the name, which was bought by BMW. VAG therefore sold Rolls-Royce to BMW and focused its efforts at its Crewe plant solely on Bentley. In 1998, VAG also bought the Lamborghini and Bugatti names — with the intention of resuming production in Italy and France, respectively. With the purchase of Cosworth Technology in 1998, VW was able to tap into the research and development for many automobile makers and leading edge work for Formula 1 racing.

Strategy 7: technological factors

There are other strategies that have enabled the company to become truly global. 'Badge engineering' means that many cars have the same components, thus reducing costs. For instance, VW Golf, Audi A3, SEAT Leon and Skoda Octavia have the same floor pan. This enables the company to meet a range of niche markets by diversifying a standard car rather than developing globally identical cars — which is the Toyota and Ford strategy. Parts are moved between the group's suppliers on a 'just in time' basis, but with many of the components now sourced locally. This enables high labour cost assembly lines in Germany to be supplied by lower labour cost parts makers in Poland, Hungary and Slovakia.

The newer plants often make more than one of the group's products. For instance, Changchung in China makes both Audi and VW cars, besides subassemblies for the plant in Shanghai. A further strategy is to combine with other manufacturers in a single plant. An assembly line at Setubal in Portugal makes VW, Ford and SEAT versions of people carriers.

VAG's strategy contrasts with other multinational car makers. Its expansion has been based on locations in a variety of markets and especially on grasping the opportunities provided by major political changes — for example, the break-up of the Soviet bloc and the opening up of China. It is a multibrand company, using technological innovations and common parts to produce a wide range of products that appeal to buyers in the developed world, while targeting production in the new markets on a more limited, cost-effective range of products.

Table 7 *The development of the VW group*

Date	Location	Region	Number of workers, 2005
1938	Founded in Wolfsburg	Germany and west Europe	105 000
1938	Braunschweig (components)	Germany and west Europe	6 600
1949	Nationalisation	–	–
1953	São Paulo, Brazil (assembly)	Developing world	21 900
1956	Hanover (commercial vehicles)	Germany and west Europe	15 000
1956	Acquires assembly plant at Uitenhage, South Africa	Developing world	6 500
1958	Kassel (supplies gearboxes worldwide)	Germany and west Europe	15 300
1964	Emden (export assembly plant)	Germany and west Europe	9 500
1965	Purchases Auto Union (Ingolstadt) from Daimler Benz, becomes Audi, merges with NSU (Neckarsulm)	Germany and west Europe	44 700
1967	Puebla, Mexico, Beetle for US market (Audi)	Developing world	15 100
1970	Salzgitter plant opens	Germany and west Europe	7 200
1972	Sarajevo, Yugoslavia, opens	East Europe	–
1976	Brussels, assembles VW and later SEAT	Germany and west Europe	4 900
1981	Acquires 66% of Chrysler, Brazil, to make large commercial vehicles (São Paulo)	Developing world	–
1982	Cooperation agreement with SEAT	Germany and west Europe	–
1982	Shanghai VW, largest car factory in China	Developing world	12 300
1985	VW Changchung, VW and Audi	Developing world	8 900
1986	Take over — SEAT (Barcelona and Pamplona)	Germany and west Europe	4 260
1989	**German reunification and break-up of the Soviet bloc**	–	–
1989	Matorell plant (SEAT)	Germany and west Europe	16 300
1990	Take over — Skoda (Mlada Boleslav, Kvasing and Vrchlabi, Czech Republic)	East Europe	23 500
1990	Take over — Skoda (Bratislava, Slovakia)	East Europe	8 200
1990	Take over — former East German Trabant plants (Mosel and Chemnitz)	East Europe	7 350
1991	Joint plant with Ford and SEAT at Setubal, Portugal	Germany and west Europe	2 800
1991	Poznan, Poland	East Europe	6 850
1992	Sarajevo plant closed due to war	East Europe	–
1993	Audi TT and VW engines at Gyor, Hungary	East Europe	5 000
1995	VW and SEAT plant Palmela, Portugal	Germany and west Europe	400
1995	Cordoba and Pancheco, Argentina; São Bernardo do Campo, Brazil	Developing world	3 100
1995	Poznan, Poland, with Skoda	East Europe	–
1998	Buys Rolls-Royce/Bentley (Crewe, UK)	Germany and west Europe	3 900
1998	Buys Bugatti (Molsheim, France)	Germany and west Europe	–
1998	Buys Lamborghini (Sant Agata, Bolognese, Italy)	Germany and west Europe	680
1998	Buys Cosworth Technology (Northampton, Worcester and Wellingborough, UK) and Cosworth Novi (Michigan, USA)	Germany and west Europe	830 / 230 (US)
1999	Curtiba, Brazil, opens	Developing world	–
1999	Engine plant, Polkowice, Poland	East Europe	980
2000	Stake in Scania, Sweden (2006 takeover bid)	Germany and west Europe	–
2000	Skoda plant in Aurangabad, India	Developing world	350
2001	Sarajevo, Bosnia Herzegovina reopens	East Europe	320
2002	SEAT part of Audi brand group	Germany and west Europe	–
2003	First cars from Bugatti and Lamborghini	Germany and west Europe	–

The location factors that VW illustrates are:

- *the role of markets*
- *political aspects of decisions about location*
- *using government aid to assist in establishing new plants*
- *the role of the global region of origin in expansion (in this case, Europe)*
- *the role of transport between manufacturing sites*
- *the importance of boardroom decisions that result in takeovers and mergers*

There is a clear sequence of development into a global company (Table 7).

This case study demonstrates the impact that new investments can have on employment opportunities in new industrial countries, as the employment data in Table 7 show. The search for lower labour costs is not without consequences, not least for wages in Germany. The threat of moving production away from Germany has impacted on wage demands and created labour unrest as workers see their living standards in relative decline. 20 000 job losses were proposed in 2006.

More detail can be found at **www.vw-personal.de**.

6

Using case studies

Question

(a) **Using a world map and Table 7, produce an annotated map to show the globalised nature of VW's operations.**

(b) (i) **With reference to a named TNC, discuss the factors that led to its development as a TNC.**

   (ii) **Assess the advantages and disadvantages of the company's operation as a TNC.**

Guidance

A checklist of possible points is provided here as a framework.

Advantages

- TNCs introduce new capital, technology, expertise and skills to the country/region.
- The new industries directly benefit the host country. The country may not have had the necessary technology or capital beforehand to develop its energy or mineral resources.
- Improvements in infrastructure, especially transport, benefit local industries and people.
- Jobs are created, especially in labour-intensive, light manufacturing.
- Exports are increased.
- The **multiplier effect** can stimulate further economic growth.

Disadvantages

- Capital-intensive investment creates few jobs.
- Low wages lead to the impression that the workforce is being exploited.
- Avoidance of local taxes or tax concessions and the export of profit mean the country gains little economic benefit.
- Most products are for export and are liable to changes in world demand, fashion and prices.
- Local resources are exploited and production is not based on local needs but on its earning potential in a world market.
- Less stringent health and safety and pollution controls mean that the workforce and local environment may become adversely affected.

The impact of globalisation and global shift

The losers and winners as a result of the combined processes of **globalisation** and **global shift** are shown in Figure 17. Globalising TNCs have the opportunity to choose locations for their branch plants that they perceive to be most advantageous. Although costs are not everything, they are a prime consideration in any **labour-intensive industry** — whether service or manufacturing.

Some locations in MEDCs are less favoured, unless there are other special attractions — such as the quality of the workforce, the beauty of the environment, accessibility, or strong support and financial inducements. Such regions are the losers.

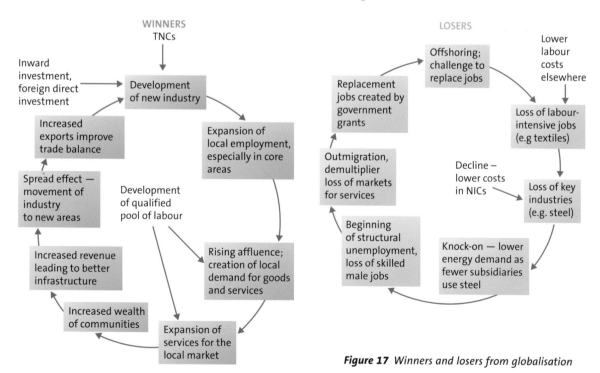

Figure 17 Winners and losers from globalisation

They face a fight to replace lost jobs with new jobs in order to retain economic stability. *Case study 3* on Saarland, a traditional coalfield location, explores how this small region exploits its full potential in an uphill struggle to diversify and re-image. A further issue for Saarland is the collapse of coal as a main energy source.

There are also winners from globalisation. A global shift of manufacturing and, more recently services, to NICs and RICs has brought opportunities for employment and economic development to many parts of the world, especially south and southeast Asia. *Case study 4* on the rise of Malaysia (pp. 34–38) explores how this has happened and evaluates the costs and benefits for an NIC. There are economic benefits, but social and especially environmental costs.

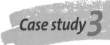

SAARLAND, GERMANY

The Saarland is one of the federal states (Länder) of Germany, forming an industrial region at the heart of western Europe. Like many other coalfield industrial regions, it has been subjected to **deindustrialisation**. Deindustrialisation is the decline of regionally important manufacturing industries due to changing demand, foreign competition and the growth of the service sector. The process has taken 50 years and reaction to the loss of key economic activities — coal mining and iron and steel — continues into this century. Saarland has a population of 1.1 million, 33% of whom live in greater Saarbrücken. Industrial Saarland is concentrated on the Saar coalfield, a strip along the Saar valley in the south, close to the French frontier (Figure 18).

Deindustrialisation

In common with other coalfield regions, the Saarland has seen the decline of its traditional economic base. Output from coal mining has declined from 15 million tonnes from 18 mines in 1957 to 5.6 million tonnes from three mines in 2003. The mines employed 40 000 people underground in 1958 but less than 2000 in 2003 (Figure 19). Productivity was 1700 kg per man shift but this rose to 3886 kg in 2003 as mechanisation improved output. The last coking plant was closed in 1999. Output was held

*Figure 18
Location of the
Saarland*

Contemporary Case Studies

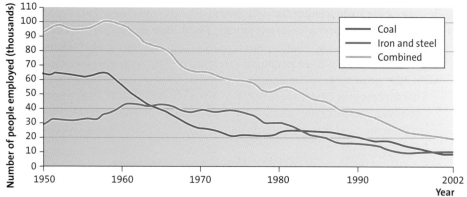

Figure 19 *Total employment trends in coal mining and iron and steel production in the Saarland, 1950–2002*

artificially high until 1980 because, under the terms of the treaty that returned the Saarland to West Germany in 1960, the mines had to supply coal to French power stations and steel works in Lorraine. **Rationalisation** has not helped employment prospects.

Iron making originated around Neunkirchen in 1593 but the iron and steel industry developed in the mid-nineteenth century, using local coking coal and limestone and iron ore from Lorraine. The plants are small and costly to operate and have been saved by merging operations within a much larger company. Today, Saarstahl, a company owned primarily by Arcelor (see Part 5), has four production sites (see Figure 18) — Dillingen, Burbach, Neunkirchen and Völklingen (the main steel-making site). The steel industry produced only 2.7 million tonnes in 2005 and employed 5219 workers — a decline of 35 000 since 1960.

In response to the deindustrialisation, the government took measures to manage the socioeconomic problems such as **structural unemployment**. The first regional action programme assisted the worst-affected areas as the regional economy went into recession. Ford established a car plant at Saarlouis in 1966, which became the focus for the largest employment sector, automobile engineering. Associated plants, such as aluminium casting in Dillingen, were also attracted to the area. Even this sector has declined since 1990. Electrical engineering producing white goods, for example Bauknecht and Bosch in Neunkirchen, was also attracted to the area. Most new jobs were for women, so that economic restructuring did not benefit those men who had lost jobs in heavy industry and mining.

Between 1984 and 1994, 16% of all jobs were lost. Even the 42 000 jobs attracted to the area since 1960 were not reducing the levels of persistent unemployment. The economy was still suffering from overseas competition. Unemployment also occurred in the service sector as computerisation replaced jobs. In 2003, tourism, banking and insurance all shed labour. The **labour pool** of the region included 46 000 unemployed (10% of the workforce in 2003).

Government response

How has the government responded to deindustrialisation of the Saarland in 2005?

- **Perceived locational advantages** — 50% of the EU's **GNP** is generated within 500 km. Planners stress Saarland's centrality in Europe, enabling it to develop into a European business hub with infrastructural links by autobahn and rail.
- **By promoting location** at the heart of Saar-Lor-Lux international region and by working across borders.

- **By developing leading industries** with potential to grow and capacity to agglomerate.
- **By creating a research and development (R&D) base** using qualified labour and research output of local universities at Saarbrücken, and Nancy and Metz in France. One thousand scientists are employed in research institutes in the new 'science strip' within the University of Saarland. Examples are:
    - German Research Centre for Artificial Intelligence (1988), Max Planck Institute for Computer Science (1990) and a new IT academy, all in Saarbrücken
    - Institute for New Materials (NanoBioNet) (1990) in Saarbrücken, working in the field of chemical- and bio-nanotechnology
    - Fraunhofer Institute for Biomedical Engineering in Homburg
    - Centre for Environmental Research
    - Institute for Environmentally Compatible Process Technology
- **By attracting foreign R&D** — the Korean Institute of Technology (KIST) has a European branch in Saarbrücken.
- **Assistance from Länder, federal and EU governments** of up to 18% of any investment is available. Further grants are given for software purchasing and substantial increases in local employment. The European Regional Development Fund (ERDF) provides €171 million for infrastructure, phasing out of old industrial areas and creating alternative employment to coal and steel. Saarland is an EU Objective 2 region, where funds are available for 'revitalising areas facing structural difficulties'. Objective 2 aims to revitalise all areas facing structural difficulties, whether industrial (such as the Saar coalfield), rural (such as north Saarland) or urban (Völklingen). Objective 2 regions often have levels of development close to the EU average but are faced with different types of economic difficulty, such as the loss of particular industrial sectors and a decline in traditional employment in both urban and rural areas.
- **By encouraging electronic business processes** — part of the Saarland's state funding programme, 'Entry into the Technological Age'.
- **Call centres** — 40 call centres have been established, employing 5000 people, mostly in Saarbrücken. These focus on online retailing (Lands End), insurance, telebanking, credit card processing, telecommunications and internet providers (AOL). Call centre location choice is backed up by EURANCE (the European Location Advisory Alliance), which advises companies about European locations including the Saarland. The multilingual skills (French, German and English) of the population were an added factor for some call centres. Saarland permits call centres to operate 24/7 and women are allowed to work night shifts, unlike in many other German Länder.
- **By promoting tourism** — 20 000 people are already employed in tourism, especially cross-border tourism, for example along the Mosel valley bordering Luxembourg. The 'Route of Industrial Culture' from Carreau Wendel, France, to Völklingen is a further proposal. UNESCO has recognised the importance that the steel industry had in the area, and designated the Völklingen ironworks a World Heritage site. The ironworks also doubles as a convention hall, art gallery and exhibition venue.
- **Starter units**, called founder centres, for new enterprises are supported by government and the academic community.

The labour pool includes 26 000 commuters from France. Labour costs are lower than in other parts of Germany. Land costs are half those of Bavaria and office rentals are one-third of those in the Frankfurt region.

In 2004, the employment profile of the Saarland's was:
- agriculture 1%
- industry 32%
- services 43%
- transport and trade 24%

This profile masks several issues:
- Employment has shifted to services, which may not be able to recruit people who used to work in heavy industry.
- New employment is increasingly dominated by women.
- The new knowledge and information economy attracts young workers, but fewer middle-aged ones.
- Over 45 years of assistance has still not met with total success. As in many other deindustrialised areas, branch plants are closing and finding replacements requires energy and effort.
- Despite being central in western Europe, the region is peripheral to Germany.
- More deserving areas for assistance (in terms of need) have lower labour costs, such as the former German Democratic Republic (which became part of Germany in 1989) and the former Eastern bloc countries that joined the EU in 2004.

The results of deindustrialisation in the Saarland are mixed. There are successes but there are also issues that remain stubbornly present, such as unemployment of unskilled males. This area can be compared to other deindustrialised areas in the UK, such as south Wales and the northeast, Sambre-Meuse (Belgium) and Nord and Lorraine (France), as well as the much larger Ruhr region in Germany.

This case study records the process of deindustrialisation, especially in coal and iron and steel (classic heavy industries), and the role of globalisation and global shift. The decline in coal as a source of energy is clearly a key factor in destabilising the traditional economy.
- *Review the 'plus features' of Saarland that mean it could deliver promising growth of new industry.*
- *Consider whether this 'new for old' has been a success. What problems remain?*

7 Using case studies

Question
(a) What are the causes of deindustrialisation? (10 marks)
(b) Examine the impact of deindustrialisation on a region. (15 marks)

Guidance
- Marks are a guide to the time you should spend on each part of a question. If you have 40 minutes for these questions, then at least 15 minutes (including planning time) should be spent on part (a) and up to 25 minutes on part (b).
- Make sure that you define the term 'deindustrialisation'.
- 'Causes' should be a list of supported reasons, which could draw on many regions affected by deindustrialisation. Make sure that you can use more than just the Saar as an example.
- The 'impact' includes closures of factories, unemployment and depression, as well as the reaction of governments at the local, regional and national level.

NICs

The term 'newly industrialising countries' (NICs) is used by economists, geographers and political scientists, and first appeared over 40 years ago. An alternative, **NIEs (newly industrialising economies)** is preferred by Malaysia. The term refers to those countries where industrial production has grown sufficiently for it to be a major source of national income. The list of countries that can be called NICs has grown over the years. Among the first generation NICs were Hong Kong, Taiwan, South Korea and Singapore — known as the 'four tigers'. Mexico and Brazil are the two Latin American NICs, and some add Argentina. Many Asian countries have tried to emulate the four original tigers. These **Asian tigers**, include Thailand, the Philippines and Malaysia.

As more countries tried to emulate the first-generation NICs, the term 'recently industrialising country' (RIC) was coined for them. Vietnam, Indonesia and Chile are RICs, although the most notable are two of the world's largest countries — China and India.

Table 8 shows some of the characteristics of a selection of Asian NICs and RICs. Some comparative data for the UK have been included. The data for Taiwan are limited, partly because it is not recognised as a separate country by the World Bank due to the continuing claims to its territory by China. Hong Kong was a British colony, but since 1997 it has been a separate territory of China.

A busy international container port

	Hong Kong	Singa-pore	South Korea	Taiwan	Malaysia	Thailand	China	Indonesia	India	UK
'Take off' date[a]	1965	1965	1962	1958	1965	1980	1978	1990	1980	c. 1800
NIC or RIC	NIC	NIC	NIC	NIC	NIC	RIC	RIC	RIC	RIC	–
Population, millions (2001)	6.8	4.35	47.6	22.75	24.3	64.6	1280.4	211.7	1065.1	59.2
GNI, $ per capita, 2003	25 430	21 230	12 020	12 941	3 780	2 490	1 100	810	530	28 350
GDP growth, 2003 (%)	3.3	1.1	3.1	3.2	4.6	6.4	8.2	4.1	8.3 (6.5% in 2004)	1.4
FDI, as % of GDP	ND	11.7	1.0	ND	5.8	0.8	4.7	2.1	0.6	0
FDI, $million, 2002	12 794	6 097	ND	ND	3 203	900	49 308	1 513	1 463	ND
Industrial growth, 1990–2002 (%)	ND	6.7	5.6	ND	6.2	3.7	5.9	3.6	6.5	1.2
Employment in agriculture (% M/F)	0/0	0/0	9/12	8	21/14	50/48	ND	57/54	ND	2/1
Employment in industry (% M/F)	27/10	31/18	34/19	35	34/29	20/17	ND	13/13	ND	36/11
Employment in services (% M/F)	73/90	69/81	57/70	57	45/57	30/35	ND	29/33	ND	62/88
Aid, $ per capita	1	2	2	ND	4	5	1	6	1	0
Aid as % of GNI	0	0	0	ND	0.1	0.2	0.1	0.8	0.3	0
PCs per thousand population, 2002	422.0	622.0	555.8	ND	146.8	39.8	27.6	11.9	7.2	405.7
Internet users per thousand population, 2002	430	504	552	883	320	78	46	38	16	423
Inward invest-ment to China, $billion, 2003	222.6	23.5	19.7	36.5	ND	ND	ND	ND	ND	11.5

[a] The 'take off' date is the date commonly accepted by economists for the initial take off into a modern industrial economy.
Source: World Development Report 2004, World Bank

Table 8 *Statistical indicators for NICs and RICs*

There are several reasons for the economic successes of the NICs:

- The role of intervention by the 'developmental state' and its leaders, who see economic development as essential for national identity. Some of the strongest cases of state intervention are in countries that are following a capitalist route to development. The state supports the domestic manufacturing sector through the establishment of strong planning bodies and through the use of subsidies for developing industries, tariffs, pricing policies and 'juggled' (state control of) exchange rates to keep exports competitive. This path was followed by Singapore and Malaysia and is usually termed 'export-led growth'.
- Support for the development of huge corporations that can control a large range of industries, such as the South Korean **chaebols**.
- The development of a strategy for industrialisation and planning the shift from labour-intensive, **import substitution** industrialisation to capital-intensive, **export-oriented industrialisation**. This is the model followed by Taiwan.

- By retaining the level of state autonomy — the degree to which the state can introduce policies independent of the vested interests of the factory owners, the trades unions (strikes are illegal in Taiwan) and even the military, who can oppose development. The state also needs to be relatively independent of institutions such as the World Bank, the International Monetary Fund (IMF) and the United Nations (UN).

It is clear that not all NICs have followed the same pathway.

Case study 4   MALAYSIA: A SECOND-GENERATION NIC

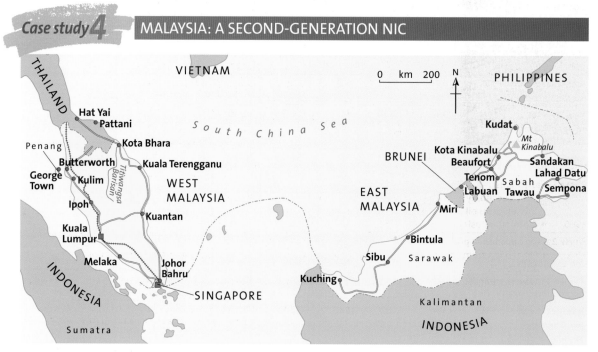

Figure 20 *Malaysia and the surrounding area*

Figure 21 shows the growth model followed by Malaysia. It looks similar to the Rostow model, but It must be stressed that the new model is very different:

- the time scale is 25–40 years and not 100 (Table 9)
- it is missing a 'basic industry phase', such as that found in the UK in the nineteenth century

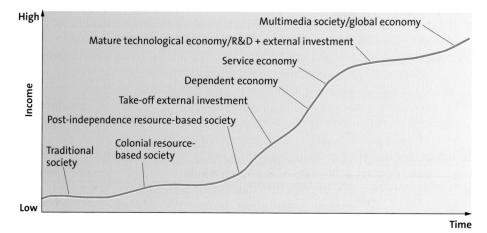

Figure 21 *The NIC economic growth path for Malaysia*

Contemporary Case Studies

Tea plantation,
Cameron Highlands,
Malaysia

Alfred Molon Photo Galleries

Table 9 Malaysia's
race to develop

Development phases	Factors
Phase 1: Early industrialisation	
1956 • 1950s economy geared to export of primary products — rubber, palm oil and timber together with tin and some oil 1965 • Clothing industry developed as import substitution	• Bringing the Green Revolution to Malaysia
Phase 2a: New economic policy (1971–90)	

1970   • Oil exports from South China Sea to Japan, in particular led growth
     • Government, led by Dr Mahatir, created central planning to enable growth of export-oriented industries from economic priority zones (EPZs), free investment zones (FIZs) in Bayan Lepas, Penang (1972) and free trade zones (FTZs)
     • Manufacturing output grew by 26% with 60% foreign owned
     • Industrial estates created, e.g. Penang and Shah Alam
     • Process state-led and not laissez-faire as in Rostow model
	• Foreign-based investment because no home-based capitalist class
	• Cheap labour – electronics workers' wages 11% of those in USA, 19% of those in Japan, 42% of those in Singapore and 71% of those in South Korea
	• Investment from global companies, e.g. Hewlett Packard, Intel, Sony, JVC
	• Neocolonial investments, e.g. ICI and Unilever
	• Management of electronics 93% expatriate in 1980s

Phase 2b: Heavy industrialisation	
1980 • More processing of oil and gas prior to export. • Growth of car and motorcycle assembly (Proton at Shah Alam, 1985, and Yamaha, 1983)	• Creating government-run industries, e.g. Petronas
Phase 2c: Service sector	
• Tourism beginning to be developed with 'Visit Malaysia' years • Need for an educated workforce apparent, so higher education becomes a major state investment both at home and by sending students away, e.g. to UK and Australia	• Build new universities such as Universiti Sains Malaysia (Science) in Penang • Encourage overseas investment in hotels
Phase 3: Taking the leap (1990–2000)	

• Strategies to leapfrog Malaysia to developed world status by 2020
     • Move from labour-intensive to high value-added industries
     • Emphasis on high technology and increasing service sector employment, e.g. Kulim Science Park, Kedah, the MSC (multimedia super corridor) project, Labuan offshore finance centre
	• Basing development goals on cultural (Bumiputras) pride, e.g. as an Islamic country that could dominate region
	• Commonwealth (Friendly) Games 1998
	• Look East policy and 2020 Vision expressed national pride
	• Role of Chinese diminished
	• More investment from region, e.g. Hyundai (South Korea) and Advent computers (Taiwan)

Economy & Development

What are the challenges for Malaysia?

■ Malaysia still relies on inward investment and state investment, and is a **dependent economy**. There are over 360 Japanese companies in the country and over 60 Singaporean companies in Johor.

■ East Malaysia, comprising the states of Sabah and Sarawak on the island of Kalimantan (Borneo), is remote and lacks the investment levels of West Malaysia. This area relies on primary products — timber and palm oil — but this reliance is denuding the natural vegetation and threatening tourist resources (such as the habitat of orang-utans). Another threat, that of the vast Bakun Dam HEP project, has been deferred after protests from indigenous tribes and environmentalists.

■ Pressures to establish a more fundamentalist Islamic state are latent at present but are seen as a threat by the Chinese and Indian communities.

■ There are labour shortages in the most-developed areas, such as Penang and Kuala Lumpur. As a result, labour costs are rising and TNCs transfer production to lower-cost countries (RICs).

■ Environmental degradation is occurring from forest clearance and the 'haze' — smoke from clearance burning on Kalimantan and neighbouring Sumatra.

■ Clearance of mangroves for commercial farming of shrimp and prawns is also degrading the environment.

■ The obsession to be better than Singapore drives policy. Developments such as the multimedia super corridors are creating a strong economic heart at the expense of the more remote regions.

Thirty years of growth: Penang

Figure 22 shows the industrial parks of Penang. The first park opened in 1972 and the last, Technoplex, in 1996. The parks house over 750 companies and employ approximately 200 000 workers. Several parks include free investment zones — areas where tax-free regimes have been a major incentive to investors. Most industries were assembly lines for major TNCs.

The problems for Penang are that the hilly island has run out of space for development, and that there is only a single bridge to the port of Butterworth. The mainland area is congested because the area is a victim of 30 years of success. There is a major tourist area in the north of the island, focusing on Batu Feringghi, and in the colonial city of Georgetown. Further industrial development might harm tourism. Already, one favourite tourist spot, the Snake Temple, is in the middle of an industrial area (near Bayan Lepas airport). Higher labour costs are forcing companies to consider moving production to Vietnam and China.

Kulim Technology Park

The 'science city of the future', covering 1440 ha at Kulim, was launched in 1996. It is located close to the late-twentieth-century industrial developments in Penang. At present, it houses 20 companies, including Intel, which employs 2500 people making server boards and motherboards for computers (see Part 3). Two wafer makers are located here, attracted by a high-quality water supply, power and telecommunications. It has an R&D zone, which mainly services the park, although private companies may locate here. There are approximately 9000 homes in the park, which also includes a golf course and a country club.

The park aims to target 'high-tech' companies with a series of incentives such as long land leases, no corporation tax for 5 years, investment allowances and flexible

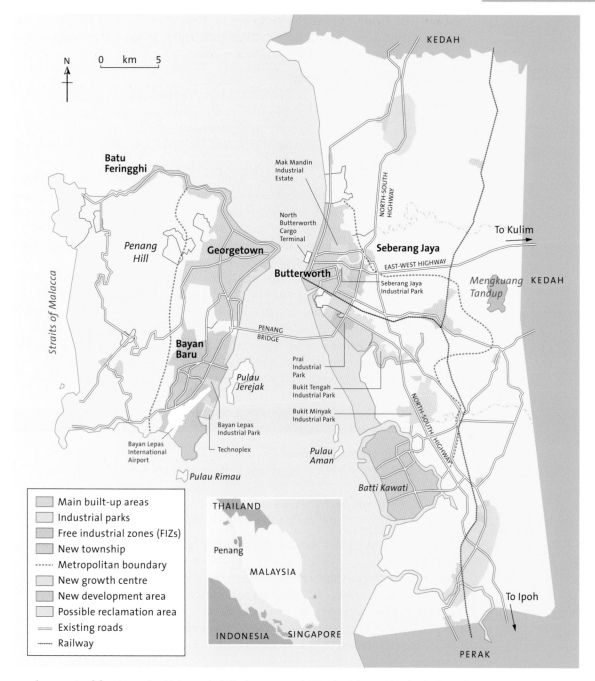

Map labels:

KEDAH

N   0 km 5

Batu
Feringghi

Mak Mandin
Industrial
Estate

North
Butterworth
Cargo
Terminal

NORTH-SOUTH HIGHWAY

To Kulim

Penang
Hill

Georgetown

Seberang Jaya

EAST-WEST HIGHWAY

Butterworth

Seberang Jaya
Industrial Park

Mengkuang   KEDAH
Tandup

Straits of Malacca

PENANG
BRIDGE

Bayan
Baru

Prai
Industrial
Park

Pulau
Jerejak

Bukit Tengah
Industrial Park

Bukit Minyak
Industrial Park

NORTH-SOUTH HIGHWAY

Bayan Lepas
Industrial Park

Pulau
Aman

Bayan Lepas
International
Airport

Technoplex

Pulau Rimau

Batti Kawati

Main built-up areas
Industrial parks
Free industrial zones (FIZs)
New township
Metropolitan boundary
New growth centre
New development area
Possible reclamation area
Existing roads
Railway

THAILAND

Penang

MALAYSIA

INDONESIA   SINGAPORE

To Ipoh

PERAK

employment of foreign scientists and skilled personnel. The last incentive is designed to attract foreign investment to a country that prefers not to have too many expatriate workers. Companies are also able to open foreign exchange accounts in a country where foreign accounts are banned.

In 2004, Kulim was designated a multimedia corridor and a cybercity by the government. This action, together with relaxing currency controls and foreign worker rules, suggests that this ambitious project is less successful than planned. Kulim would like to house biotechnology companies, but none has come. Why could this be so? The

Figure 22 *Penang and the surrounding area*

location (see Figure 20) on the periphery of West Malaysia and to the east of successful Penang may be a deterrent. Other competitors in Asia offer cheaper labour costs. Biotechnology is still mainly an MEDC activity.

Multimedia super corridor (MSC)

The MSC is a 50 km × 15 km, government planned, 750 km^2 corridor, equivalent to the area of Singapore. The MSC stretches from Kuala Lumpur city centre to KLIA (the new Kuala Lumpur International Airport). It sandwiches two major new urban areas at Cyberjaya, 'the world's first intelligent city', and Putrajaya, the new capital area. The other element is a technology park.

Cyberjaya, launched in 1999, will have 28 000 'smart' broadband-connected homes by 2010, housing 120 000 people (a working population of 50 000). It is home to the Multimedia University (3000 students, mainly of IT). There are 326 mainly IT-based companies, out of 1293 companies in the city, together with 'smart schools' and a tele-medicine centre. All the normal urban services are here in this city 30 minutes from central Kuala Lumpur and 20 minutes from the airport.

Every effort has been made to gain Malaysian investment in the industries in the MSC. 73% of the companies, employing 24 000 people, are Malaysian-owned. This masks the fact that many companies (988) are subsidiaries of major TNCs, created to enable them to conform to national law. Investment into the region is mainly from other NICs and is dominated by Japanese and Singaporean companies. Of the 97 European companies, 29 are from the UK, 15 from The Netherlands and 14 from Germany. As in Kulim, biotechnology companies are not being attracted to the area.

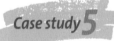

Case study 5 — INDIA AND CHINA

The awakening giants compared

The accelerating growth of the Indian and Chinese economies is a phenomenon of the twenty-first century.

- By 2025, China will be the world's largest economy.
- By 2050, India will be the world's third-largest economy.

Together, these two giant nations are currently home to one-third of the world's population.

Comparisons are inevitably made between the two countries.

- Having been classified as RICs, they are now considered to be NICs, experiencing rapid change towards becoming advanced economies.
- Both contain large populations, with emerging middle classes, leading to potentially huge internal markets — the middle class in India contains as many people as the whole of the USA!
- Both have massive demands for land, water and energy to drive development (e.g. the Three Gorges Dam, p. 41).
- Both face issues stemming from environmental degradation and decay (pollution).
- Both have made strides to reduce poverty — assuming an income of $1 a day as the threshold, currently 35% in India and 17% in China are below poverty level.
- Both contain substantial rural populations — often impoverished subsistence farmers (70% rural in India, 60% in China).

- Both exhibit marked regional disparity between cities, such as Shanghai or Bangalore, and their impoverished rural peripheries, such as northwest China and Rajasthan respectively.
- Both have an under-reported potential HIV/AIDS pandemic, which could slow their growth.
- Both face similar challenges of unemployment, or finding jobs for the entire workforce.
- Both spend a high percentage of GDP on defence because of external disputes — China has the Taiwan issue and India has the Kashmir issue with Pakistan.

In spite of these comparisons, the means of economic progress are fundamentally different (Table 10). India follows a democratic route, yet China remains a communist country but wedded to capitalist economic growth. India sees core growth around Indian-developed service industries. China concentrates more on manufacturing fuelled by foreign direct investment (**FDI**) from Japan, the USA and Europe.

The importance of the two countries is shown clearly by the data in Figure 23. It has been estimated that, in 1820, China generated 33% of the world's output and India generated 16%. Throughout the nineteenth century and up to 1950, this share declined as Europe and the USA became the core areas of industrial production. The low

Table 10
Contrasting India and China: economic profiles

	China	India
Infrastructure	• Large-scale investment — eight times more spending than India	• Investment is low other than IT
Labour force	• Increasingly deregulated but fewer entering the labour market as a result of the one-child policy dividend	• Highly regulated and growing number entering the labour market until 2050, called the demographic
Industrialisation	• Trade in goods 49% of GDP • 5.8% of world merchandise exports in 2003 (fourth in the world) • Growing at 12.6% per year	• Rate is slow • Trade in goods 21% of GDP • 0.7% of world merchandise exports (31st in the world) • Growing at 6.0% per year
Steel industry	• Output: 250 million tonnes in 2004	• Output: 40 million tonnes (80 million tonnes in 2008) • Posco (South Korea) to build a plant in Orissa
Service sector	• Services exports 2.6% of the global total (ninth in the world) • Growing at 8.6% per year	• Services exports 1.4% of the global total (21st in the world) • Growing at 7.9% per year
Tariffs	• Low, averaging 13%	• Protected industry, averaging 28%
Education	• Universal basic education • 98% of children have 5 years of primary education • 6% illiterate	• Education still an elite activity • 47% have 5 years of primary education • 39% illiterate
Savings	• High: 44% of national income	• Low: 22% of national income
Underlying philosophy	• The state will develop the economy for the people	• The state will develop the economy for its clients, be they Indian or from overseas
Energy	• Consumption of oil is rising rapidly • Biggest consumer after USA	• Consumption of imported oil is rising rapidly • Competing for the same imported resources • Sixth largest emitter of carbon dioxide

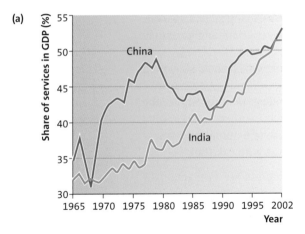

(a)

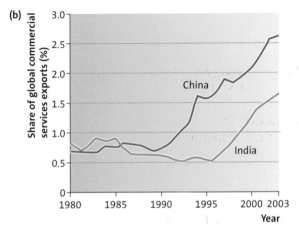

(b)

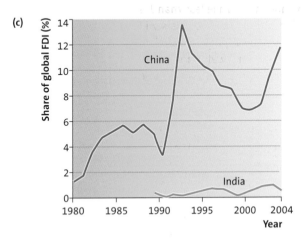

(c)

point for China and India was 1950, when their shares of output were 5% and 3% respectively. Since 1970, this position has begun to change. China was locked into a rigidly state-controlled socialist political system at the very time that the Asian tigers were beginning their path to development. India was also a country of autocratic one-party control, emerging from over a century of colonial rule.

Since 1980, China's economy has grown at an average of 9.5% per annum, while India's growth has averaged 5.7%. China's GDP has been rising faster than any other country in the world and India's was ninth fastest between 1980 and 2003. Their economies have grown at almost the same rate as did those of Japan (between 1950 and 1973) and South Korea (between 1962 and 1993).

What was the trigger for this rate of economic growth? In both cases, it was the gradual movement away from state control to a market economy as a direct result of their governments realising that the economies were performing badly. Both countries were able to make considerable use of their vast labour force potential and the fact that productivity was low. They were able to offer global companies a supply of cheap, trainable labour, using more productive technologies.

China's growth has been more spectacular because it has retained centralised management of the economy. It has the ability to enforce strategies with all the force of the state. These strategies have included:

- a fixed exchange rate, so that it could buy raw materials at a competitive price and sell goods cheaply — one of the ingredients of rapid growth
- a cheap labour force
- a labour force with a reputation for hard work
- the ability to transfer workers from low productivity employment in agriculture to higher productivity jobs in manufacturing
- political stability
- a government oriented towards economic development

Figure 23 *India and China: (a) % share of services in GDP; (b) % share of global commercial services exports; (c) % share of global FDI*

China has copied the growth pattern of Japan. Even so, it will take until 2030 for China to have the same per capita GDP as Japan. In 2005, GDP in China was the same as Japan in 1961 and the same as South Korea in 1982.

In contrast, India's growth has taken place in a country that prides itself on its democracy and diversity. India's initial growth was largely in the service sector, which does not create many jobs, yet produces high returns. In particular, IT jobs created to recruit new graduates and expatriates returning to India are booming. IT now contributes 3% of India's GDP.

Bangalore is the centre of the Indian software industry. Texas Instruments began testing chips in Bangalore in 1985. Since then, other companies, such as Hewlett Packard and General Electric, have established laboratories. In 2004, Google opened an office in Bangalore and was joined by Microsoft in 2005. These companies were attracted by the qualified labour force, and up to 60% of their employees have higher degrees. In 1985, there were 150 000 software engineers in Bangalore — numbers that rival Silicon Valley in California.

Another growth area is biotechnology and drug manufacturing, with major pharmaceutical producers opening R&D laboratories, for example in Mumbai. Researchers in India cost 33% less than similar workers in the USA and Europe, while, because of poverty, people are more willing to enrol to test new products.

Current challenges for India and China
India
Reducing poverty remains India's greatest challenge. India is still home to around 260–290 million poor people. These numbers rise to 390 million if poverty is measured by the UNDP standard of those living on less than US$1 a day. Rural development is essential to raise the incomes of the poor. An estimated 1 million workers leave agriculture each year, so creating non-farm employment is critical.

Improvements in transport infrastructure and in the investment climate are required to sustain the country's recent growth. Severe bottlenecks in power, water and transport impede India's economic competitiveness. Basic services, such as improved health and education, need to reach all citizens. Major changes will need to be made to ensure the effective delivery of these services, especially to the vulnerable segments of the population.

HIV/AIDS has the potential to upset much of India's recent progress. The disease is spreading quickly and placing the country in danger of a pandemic. While less than 1% of the adult population is currently estimated to be infected, the total numbers will soon be greater than any other country in the world because of India's large population.

Environmental sustainability needs to be ensured. Diminishing water resources, increasing pollution and global climate change are strategic challenges that require long-term vision.

China
The pace of economic development in China is having a severe effect on the environment in some areas. The Three Gorges Dam on the Yangtze River, the world's largest HEP project, will supply 10% of China's electricity. Construction started in 1994 and will be completed in 2009. It will flood an area of 630 km^2 and will create a lake 200 km long. Many people fear that the lake will fill with sewage from upstream cities, so the government is trying to build 200 sewage plants to reduce the forecast pollution. The other effect is a human one, with 720 000 people already moved from the area that will be flooded (Figure 24).

Low production costs are a threat to producers in MEDCs. In 2005, the EU's quota of clothing imports from China was reached and exports were left in warehouses in European ports. The quotas were introduced because Italian, French and Spanish clothing manufacturers felt threatened by Chinese production. Similar fears were voiced in 2006 about shoe production. Such trade wars could affect future industrialisation.

There are several economic challenges for China:

- If the state-controlled exchange rate is permitted to float in 2007, Chinese goods may become less competitive.
- The cost of energy is driving up global prices, which will impact on economic development in China.
- Chinese demand for steel has driven up prices, which China may not be able to afford in the future.
- Workers need to migrate to the new industrial regions, and enough housing needs to be provided for them.
- A lower birth rate and ageing population lead to demographic challenges.

Figure 24 *The Three Gorges Dam*

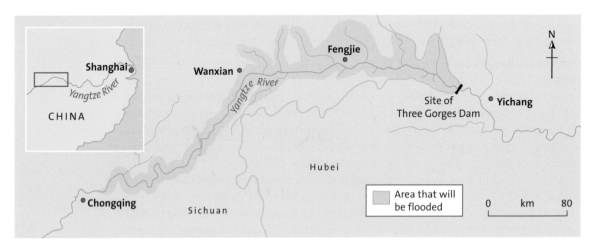

Using case studies

Question

Using the information for China and India and Figure 25:

(a) explain why China is currently more successful than India

(b) suggest why many economists feel that, by 2050, India may be 'the tiger in front'

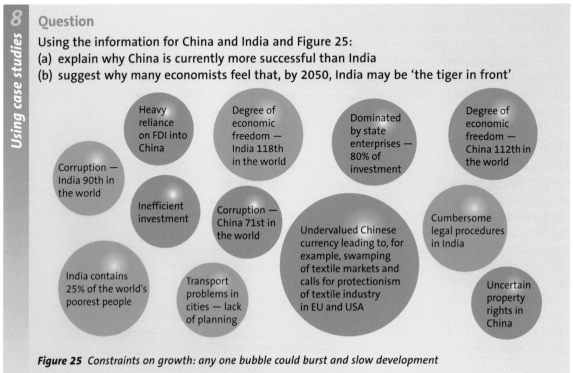

Figure 25 *Constraints on growth: any one bubble could burst and slow development*

Contemporary Case Studies

Changing locations

In this section, we look at how the key locational factors shown in Figure 26 have influenced changes in the location of steel manufacture — a key world manufacturing industry. We then focus on the present location of the Rolls-Royce car company and its choice of Goodwood in Sussex as the site for its factory to serve a global market. This case study shows how factors determining the choice of location have changed over time.

Figure 26 *Key factors in choosing an industrial location*

The steel industry

Iron and steel production illustrates how the 'traditional' location factors have changed over the past 200 years. Present-day patterns of iron and steel production can be explained by the influence of new factors related to the changing global economy.

Traditional location factors

Iron and steel production has often been used to illustrate Weber's principles of industrial location. Weber's theory is based on the 'least-cost location'. He developed the concept of a material index to calculate the amount of weight lost or gained in the production process. The growth and evolution of the iron and steel industry in the UK and Germany illustrate his principles.

Early production
Iron making commenced where there was iron ore and where the power source — wood for charcoal — was plentiful. The Weald, in Surrey and Sussex, and the Forest of Dean, in Gloucestershire, were two areas where iron production was important in the sixteenth century. In Germany, the hills of the Sauerland were another example.

The Industrial Revolution

Production shifted to the coalfields following the introduction of the use of coke (instead of charcoal). This was because of a favourable weight ratio: 8 tonnes of coke and 4 tonnes of iron ore produce 1 tonne of pig iron. In addition, the iron ore came from the coal seams (blackband ores). Therefore, locations such as Ebbw Vale and Consett in the UK, Essen and Dortmund in the Ruhr (Germany) and Völklingen in the Saarland (Germany) grew rapidly. The industry was oriented to the heaviest raw material — coal — which has the greatest weight loss during the manufacturing process.

Late-nineteenth century

Further new locations arose once technologies had been developed to exploit the low-grade iron ores in Jurassic limestone. Works were opened in the UK in Scunthorpe, Corby and Teesside to exploit these ores in exactly the same way that the ores around Salzgitter and in Lorraine were exploited by German companies. It was now the case that the raw material with the greatest weight and weight loss was the ore. A few plants were located on high-grade ores, such as Workington, which used haematite ores of West Cumbria and the local coalfield.

Coastal shift

By the mid-twentieth century, most of the early ore supplies in Europe were exhausted and only low-grade supplies remained. Steel makers were increasingly turning to the high grade (65% +) ores imported from Norway, Canada and further afield. Bulk cargo transport costs had fallen and new mining and manufacturing technologies aided change. As a consequence, **break of bulk** integrated steelworks developed on the coast — for example at Port Talbot, Llanwern and Teesside in the UK. In Germany, emphasis moved to the mouth of the Ruhr at Duisburg, where large barges could bring ores up the River Rhine from Europoort (Rotterdam) in The Netherlands.

Inertia

As a result of the industrial depression of the 1930s, when demand fell and recession had reduced traditional industries such as steel making, government intervention in location decisions became a new factor. In the UK, this was combined with nationalisation of the industry, so that location decisions were made for strategic regional planning and political reasons as much as for finding the cheapest location. As a consequence, government subsidies created forces of **inertia**, which determined that uneconomic plants, such as Ebbw Vale, were retained for social reasons as sources of employment in declining coalfield regions. In Germany, similar support was given to steel makers in the Saarland after 1960. Some locations, such as Sheffield, used scrap iron as the main raw material for electric arc furnaces to make specialised steel, in spite of its inland location.

Economies of scale

By 1975–90, coalfield locations were a throwback to the past or anachronistic. New steel-making technologies demanded little coke and used high-grade imported ores. Old, small plants were closed or rationalised to process steel, and the new, larger-scale, integrated plants on the coast were the most cost-effective. Companies merged to gain **economies of scale** — for example, ARBED in Belgium, Luxembourg and France, Krupp-Thyssen in Germany and British Steel. Government regional planning strategies still determined new plant siting — for example, Taranto in the Mezzogiorno, Italy, and Dunkerque in northern France.

Global competition

After 1970, the signs of competition from new, large-scale, technologically efficient and low-labour-cost plants increased as works expanded in Japan and were developed in the nascent NICs such as South Korea (an early sign of global shift). Many new steel makers were state sponsored, although most were private rather than nationalised companies. Steel could be made more cheaply, despite importing all the raw materials, and could be sold across the world at prices that competed with the older makers. The reaction to this was further rationalisation of steel makers into larger groups and the closure of the smaller inefficient plants. In 2004, Corus, which had taken over British Steel, reduced capacity at its plants in the UK by 20%. To survive global competition, many small steel mills specialised. For instance, Salzgitter in Germany focused on tubes and pipes.

Since 1945, a highly protected steel-making industry had grown up in the Soviet bloc. Stalinist ideology dictated that every country in the bloc should develop its own iron and steel industry because the workers in such an industry formed the proletariat, the social foundation of the workers' states. Much of the industry was located on the main sources of raw materials, for example in southern Poland and Kazakhstan.

---

9 Using case studies

Question

Using a blank map of the UK, produce an annotated map to show the main stages in the changing location of iron and steel manufacture outlined on pages 43–44.

Guidance

Colour-code your locations according to their stage in the locational evolution of the industry.

Twenty-first century location

The origins of the changes that have given rise to the location of steel making today are complex (Figure 27). Some economists cite the global political shift to the right during the Reagan and Thatcher years (the 1980s), which made the role of the market very important. At the same time, the collapse of the Soviet bloc after 1989 led to the privatisation of production in the former member states. Investment policies in the new companies, all of which were seeking

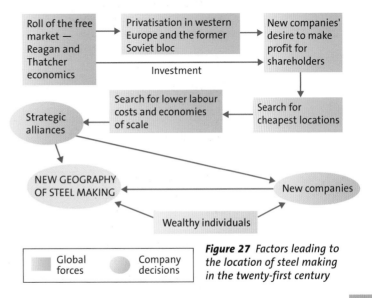

Figure 27 Factors leading to the location of steel making in the twenty-first century

profit, led to a set of policies that have come to impact on the current global pattern of steel making. The impact of the free-trade initiatives promoted by the WTO and GATT laid steel makers open to the forces of globalisation and the formation of global companies. Steel accounts for 0.5% of global GDP and, if its downstream uses by industries such as automobile assembly and construction are included, it contributes to 20% of global GDP. This explains its designation as a 'keystone industry' and the desire of many emergent industrial nations to produce their own steel.

Figure 28 shows world steel production in 2005. Production has shifted to Asia, which accounted for 50% of production in 2005. Table 11 emphasises this change, although it does stress other characteristics, such as the recent development of some transnational, and even global, steel companies.

Figure 28 World steel production, 2005

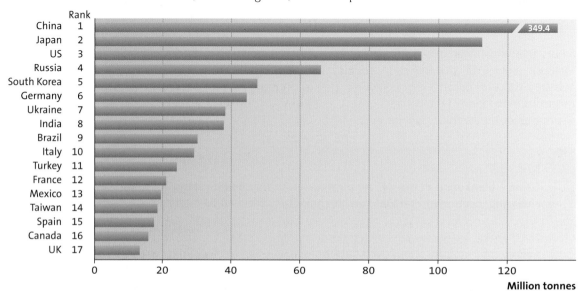

Table 11 *Major steel production companies, 2004–05*

Company name	Created from	Location of HQ	Production (million tonnes)
Mittal Steel	LNM, US Steel (US Steel/Bethlehem Steel)	The Netherlands Antilles	61
Arcelor	ARBED, USINOR	Luxembourg City	47
POSCO E&C	Part of **Chaebol**	Seoul, South Korea	32
Nippon Steel	Yawata Steel and Fuji Steel	Tokyo, Japan	32
JFE	NKK, Kawaski	Tokyo, Japan	30
Nucor	Merged with six small makers	Charlotte, USA	17
Baostahl	Created 1997	Shanghai, China	20
ThyssenKrupp	Thyssen and Krupp	Dusseldorf and Essen, Germany	20
Corus	British Steel, Koninklijke Hoogovens	London, UK and IJmuiden, The Netherlands	19
Riva	IRI	Milan, Italy	15
World			**1107 (in 2005)**

Contemporary Case Studies

Global shift in steel production

Asian companies produced 50% of output in 2005. What caused the change from production dominated by Europe and North America to production dominated by Asia and by large corporations?

- Better technology — mining of ores, transporting ores in bulk carriers and smelting technologies. Some of the major sources of ore and coal are in Brazil and Australia. Three companies now dominate 70% of iron ore mining.
- Oversupply in the twentieth century kept prices down so much that, in real terms, 2004 prices were 30% below those in the 1970s. North American and European companies were unable to compete on price due to their higher labour costs compared with China, Brazil and South Korea, even in such a **capital intensive industry** as steel. The USA tried to use a tariff barrier to protect its producers from cheap steel imports, but this was declared unlawful by the WTO following EU protests.
- Privatisation — many companies in the former Soviet bloc were government owned. After 1989, privatisation in Poland, Romania, Kazakhstan and Russia left these companies in need of investment, owners and markets.
- Consolidation among producers to create economies of scale and maintain a presence in the market place was a further strategy. Consolidation eliminated weaker production plants and gave the new enlarged companies access to cheaper production facilities with lower labour costs. The bigger mergers are noted in Table 11.
- Even if companies did not merge, many were taking shareholdings in other producers, perhaps with a view to subsequent takeovers. For instance, Arcelor has an interest in producers in Russia and Brazil, and in 2006 Mittal bid for Arcelor.
- The explosive growth of the Chinese economy since 1995 resulted in China consuming 25% of global steel production. Chinese output was 350 million tonnes in 2005 (up from 109 million tonnes in 1997). The consequence of Chinese demand has led to price increases, which benefit the bigger, consolidated producers most.
- MEDC companies are seeking mergers in Asia. It is economic nonsense to transport steel, which is very heavy, over long distances when profits can be made from manufacturing close to, or in, the major markets such as China.
- Chinese prices undercut everybody because the government imposed a fixed exchange rate (until 2005), which made steel cheap to export and to buy. Firms from Japan (Nippon) and Europe are seeking joint ventures in China with state-owned Baostahl.
- Environmental legislation is curbing investment in Europe. Krupp in Duisburg, Germany, is not modernising or expanding its facilities because of the cost of complying with carbon dioxide legislation.

MITTAL STEEL

Case study 6

Mittal Steel is the first truly global steel manufacturer, having grown from its base in Calcutta, India, where it was founded in 1976. Over the past 30 years, it has acquired production throughout the world, so that it has steel making or processing facilities on every continent. It is registered as a company offshore in The Netherlands Antilles and is 88% owned by the Mittal family. It employs 330 000 people worldwide and produces 113 million tonnes of steel. Table 12 lists the main companies owned by Mittal Steel in 2006. As an example of its importance, it supplies 30% of all steel used in American car assembly.

Table 12 The major acquisitions of Mittal Steel, 1976–2006

Company	Locations acquired	Year	Other information
Arcelor	France, Belgium, Luxembourg	2006	Bid to take over
Hunan Valin	China	2005	36% ownership
BH Steel	Bosnia	2004	–
Weirton	Georgetown, Trinidad	2004	–
Polska Stal	Katowice and four plants in south Poland	2004	6 million tonnes — 70% of Polish production
Siderurgica	Hunedoara, Romania	2004	0.75 million tonnes
Balkan Steel	Macedonia	2004	–
Nova Hut	Ostrava, Czech Republic	2003	Majority of Czech output of 6.5 million tonnes
Petrotub	Roman, Romania	2003	0.5 million tonnes, pipes
Bethlehem Steel	Chicago, USA	2003	–
LTV Acme Steel		2002	–
ISCOR	Vereeniging, Vanderbijlpark and Saldhana, South Africa	2002	5.3 million tonnes, eighth lowest cost producer in the world
Sidex	Galati, Romania	2001	5 million tonnes — 50% of Romanian output
Annaba	Algeria	2001	1.8 million tonnes
Trefileurope, Unimetal	Seven sites in France, one in Belgium and one in Italy	1999	Wire and nails
Inland Steel	Chicago, USA	1998	5% of US production
Stahlwerk Ruhrort	Duisburg, Germany	1997	1.3 million tonnes steel; iron from Thyssen
Ispat Walzdraht	Duisburg, Germany	1997	Wire
Ispat Unimetal	Amneville, France and Luxembourg		1.5 million tonnes, wire
Ispat Shipping	London, UK	1995	Transport of raw materials
Karmet	Temirtau, Kazakhstan	1995	5 million tonnes; 1.5 billion tonnes coal reserves
Hamburger Stahlwerk	Hamburg, Germany	1995	0.5 million tonnes, mainly for wire
Ispat Sidbec	Quebec, Canada	1994	–
Mexicana, Sibalsa	Michoacan, Mexico	1992	4.0 million tonnes
Ispat Caribbean	Couvas, Trinidad and Tobago	1989	2.7 million tonnes
Ispat Indo	Surabaya, Indonesia	1976	0.65 million tonnes

The case of Nova Hut (Czech Republic) illustrates the processes of changing a nationalised steel maker into part of a global company. Nova Hut was losing money and, with 123 000 workers, it was overmanned at the time of the collapse of the Soviet bloc. In 2003, Mittal took over the plant, immediately reduced the workforce to 8000 and rationalised production, purchasing and sales. Higher value products, which could be sold in western Europe, were developed as a result of new technologies. The EU states of eastern Europe are proving to be a successful manufacturing location, combining high labour skills with much lower labour costs than in western Europe.

10 **Question**

Using case studies

(a) Use a world map to show the global development of Mittal Steel. Use colours to show the dates of expansion.
(b) Use located bars and annotations to show the nature and amount of output.
(c) Write a commentary of around 250 words, using information on pp. 47–49 to explain your map.

ROLLS-ROYCE SITE SELECTION

Case study 7

Rolls-Royce Motor Cars opened a new assembly plant at Goodwood, just outside Chichester, West Sussex, in 2003 when the first 'new' Rolls-Royce from the plant was unveiled. The evolution of the company up to 2003 illustrates many of the factors that have influenced industrial location over the past century.

Origins of the company

The early career of Henry Royce (1863–1933) saw him working as an apprentice on the railways, at a toolmaker's in Leeds and with early electricity companies in London and Liverpool. In 1883, he established a company in Manchester that made light fittings, dynamos and cranes. By 1903, his experience with early motors made him determined to build his own cars, and the first was produced in 1904.

Charles Rolls (1877–1910) came from a wealthy family. His career path via Eton and Cambridge included achieving a world land speed record. In 1902, he began selling cars but lacked reliable vehicles to sell. In 1904, Rolls met Royce in Manchester and soon agreed to make chassis for Rolls-Royce cars. At that time, the bodywork was attached by many individual companies working to order.

Four location factors were involved.

- The origins of the company were, like other motor companies such as Morris and Opel, partly in the mechanical and electrical engineering sector and emerged by chance.
- The location of the first production plant was Derby — there was a labour force in the city with many of the skills required.
- The personalities and ideas of the two men gelled and they were able to work together — this is a behavioural factor. It also emphasises how an entrepreneur can have a major impact on location: Boots in Nottingham and Wedgwood in Stoke are two other cases.
- Bodywork was a subsequent assembly, closer to the main markets, especially in London.

In 1931, Rolls-Royce acquired Bentley and continued to produce the Bentley models. It was only in 1946 that the company decided that it should assemble the complete car. To do this, it had to divorce car assembly from the other functions at Derby, such as the increasingly important aero engine business that had grown during the Second World War. In 1946, the assembly plant was relocated to Crewe, into a factory that had been used to produce aircraft engines. In 1973, Rolls-Royce ran into severe financial trouble due to its involvement in the development of advanced, technically complex aircraft engines and the company became state owned. The car company was subsequently sold to Vickers Engineering.

Relocation to Chichester

In 1998, Rolls-Royce/Bentley was sold to Volkswagen (VAG), although the rights to the Rolls-Royce name were sold to BMW. The outcome of this action was that BMW had to find a new location for Rolls-Royce because VAG/Bentley refused to continue production of Rolls-Royce cars at Crewe after the end of 2002. A location for production that was rejected immediately was Germany. BMW's business and public relations intelligence had shown that Rolls-Royce cars had to be British-built. The company had 4 years to select a new location and begin production at that location.

The following location criteria, an example of micro-location factors, were drawn up by the company:

(1) a distinctive site in an area of natural beauty into which the factory could blend
(2) a site of at least 12 hectares to accommodate the buildings for the assembly and the headquarters office
(3) good transport links with the rest of the UK and internationally to enable delivery of parts
(4) a high-calibre workforce available in the area, with some experience in quality engineering
(5) a location with access to an on-site or nearby test track

Figure 29
The location of the Rolls-Royce site, Goodwood

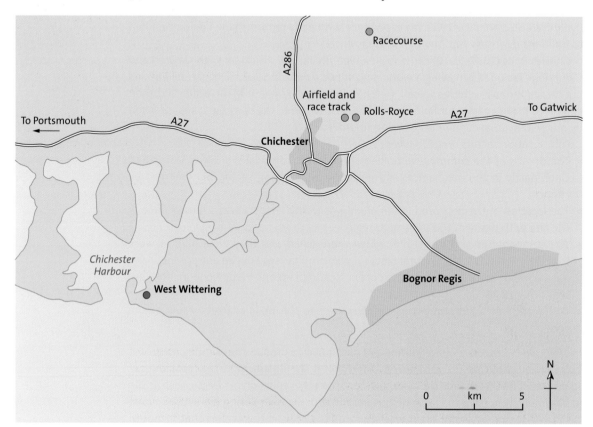

Contemporary Case Studies

(6) a location with associations with Rolls-Royce
(7) proximity to international cultural and sporting attractions where buyers might be entertained
(8) environmentally friendly in its image, with minimal local opposition

The short list of potential locations included:
- Cowley, Oxford, where another BMW product, the Mini, is built
- close to the M40, perhaps near Warwick (an area long associated with car production)
- Goodwood, West Sussex

In 2000, K.-H. Kalbfeld, Managing Director BMW/Rolls-Royce Project, said: 'The Chichester region offers everything we need for our project. After carefully analysing all the details, we have chosen Goodwood as the most suitable location'.

Table 13 shows how Chichester fitted the criteria listed above.

Where do the parts of the modern 'Roller' come from?

- The design team is still based in London.
- The engineering process team is based with BMW in Munich, Germany.
- The space frame (the basic box around which the car is assembled) is produced in a specialist unit at Dingolfing, Germany.
- The engines come from the BMW M Series facility in Munich.
- The leather hides come from Bavarian bulls, where the carcass is used for meat; this is an attempt to justify the extensive use of leather from several animals.
- The wood veneers utilise mahogany from west Africa; walnut, birdseye maple and black tulip from North America; and oak and elm from Europe.
- The veneers are shaped in the UK.

Criterion	
1	58.3 hectare site on the Goodwood estate located just outside the boundary of the proposed South Downs National Park, currently an AONB. It is close to sailing opportunities of Chichester Harbour and the Solent (Figure 29).
2	The site is set into a disused 17 ha gravel pit, so that it is partially below ground level.
3	The A27 trunk road provides access to Portsmouth and Southampton (for cross-channel ferries). Rail and road links to Heathrow and Gatwick airports are satisfactory (60–90 minutes). Goodwood has a small private airfield. It is within 1 hour's journey time of BMW's UK headquarters at Bracknell.
4	Boatyards in the area have trained, skilled crafts-people and build products to exacting standards.
5	The former Goodwood motor racing circuit is within 1 km of the plant. The Festival of Speed and the Revival Meeting attract up to 150 000 visitors.
6	The company talks of a 'spiritual connection' with Sir Henry Royce who lived and died at West Wittering, about 20 km from the site.
7	Goodwood Racecourse is within 5 km and attracts up to 25 000 racegoers for each of its 12 racing days. Chichester has its Festival Theatre.
8	Nicholas Grimshaw and Partners, the architects for the Eden Project, were chosen to design the buildings. The roof of the building, set into the gravel pit, is covered with plants and is the largest 'living roof' in Europe. This helps insulate the building and cuts heat loss. Lakes on the site contain heat exchangers to replace conventional air conditioning. As a result, electricity consumption is 20–33% of the normal for such a facility. Schedules have been drawn up to enable just 20 lorry movements per day. All wood veneers come from sustainably managed forests.

Table 13 Rolls-Royce and Chichester

Rolls-Royce's (BMW's) locational choice illustrates several trends in modern transnational companies:
- The product is assembled in the UK, but the major parts that make up the product are almost entirely German (from BMW — the parent company). Other parts come from around the world.
- Chichester was selected primarily because BMW realises that Rolls-Royce would lose its reputation if it was perceived as anything but British. BMW made a similarly successful decision with Mini production at Cowley.
- The handmade nature of assembly and the customer base, with a global market for the 1000 cars produced each year, all pointed to an assembly site that could be visited

www.bosham.org

The Rolls-Royce factory at Goodwood

by customers living the 'celebrity lifestyle'. The Chichester area has facilities to satisfy these guests — quality horse racing, sailing, a private airfield, top-class hotels and a quintessential 'English' landscape in the South Downs with attractive villages and country houses.

■ The decision was more about behavioural reactions of the market than the costs of production.

■ Local opinion was won over by outstanding attention to environmental considerations and the promise to source some materials (mainly for the office functions) from the region.

■ A skilled labour pool was important. The company selected an area with only 1% unemployment and was not interested in the unskilled. More than three people applied for each job.

■ The ability to import and export was stated, but access to ports was of secondary importance with such a high-value product.

This case study is a surprising one, as it shows how the traditional locational factors do not really apply to the final choice of site for a specialist 'niche' car business owned by a global company.

11 Using case studies

Question

(a) Use Figure 26 on page 43 to identify and rank the key locational factors in BMW's decision to choose the Goodwood site for Rolls-Royce.

(b) Using examples, show how the factors affecting the location of manufacturing industries have changed in the past hundred years.

Guidance

■ Planning — first, make a list of locational factors, as shown in Figure 26. Use two headings: 'traditional locational factors', e.g. raw materials and energy supply, and 'present-day locational factors' such as globalisation, behavioural factors and environmental considerations.

■ For (b) you can use steel to illustrate factors such as raw materials, markets, labour, break of bulk point and even inertia.

■ Use the Rolls-Royce case study to illustrate personal choice and behavioural factors.

■ Global markets could be illustrated using VW (*Case study 2*). The early stages of the development of VW illustrate political and strategic factors.

■ Make sure you can name locations to support your points.

Part 6

Service sector locations

Today, the service sector dominates the world's advanced economies — as the Clark–Fisher model shows (see Figure 3, p. 1). As people become wealthier, their consumption of travel, cultural activities and education increases, and more people begin to save and invest their money. There are two types of service activity in MEDCs: the UK/USA/Canada service economy model, where independent suppliers provide services; and the Japan/Germany model, where many manufacturing companies provide services.

In the twenty-first century, services are growing because of the increase in the volume of information and knowledge held electronically. The number of knowledge- and education-based workers is increasing rapidly, to become the largest employment group in many cities. In MEDCs, a **post-industrial society** exists, where a well-paid professional elite is itself served by a low-paid service sector of cleaners, e-commerce workers, shop assistants and front office staff, for example.

The growth of the service sector

The growth of the service sector is also known as 'tertiarisation'. Many of the factors influencing service sector growth are interlinked (Figure 30).

Workers in a call centre

Topfoto

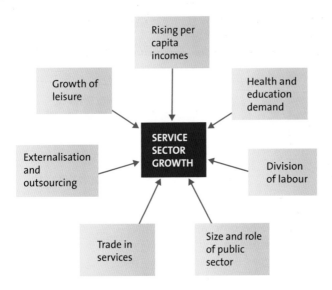

Figure 30 Factors influencing the growth of the service sector

With rising incomes, we bank, save, invest and spend more. We own goods that need maintenance (e.g. cars and washing machines), eat out more and have more holidays. We live longer and need more care in old age. We remain in education and training longer or even receive life-long education. We expect government to provide us with services, such as waste disposal and recycling, to support economic growth and to plan for the future. We expect laws to be enforced to give us the quality of life that we demand. All governmental support and intervention increases the number of public sector workers. It is estimated that about 20% of total trade is in services being bought and sold in activities such as transferring money and selling insurance. Companies divide service functions, for example by establishing a human resources department or by the employment of legal experts. Recently, large firms have begun to either buy in or outsource activities such as accounting, customer relations and routine tasks.

What are services?

- Finance, insurance and real estate (FIRE) — includes banking
- Business services — advertising, architecture, PR, accounting, R&D
- Transport and communications — including electronic media
- Wholesale and retail trade
- Tourism — including entertainment and hotels
- Government (local, regional and national) — healthcare, education, social services
- Non-profit agencies — charities, churches, museums, private healthcare
- Personal services — such as domestic cleaners, child minders, window cleaners

Case study 8

CREDIT SUISSE FIRST BOSTON (CSFB) GLOBAL SERVICE ACTIVITY

Credit Suisse was founded in Zurich in 1856 to serve affluent clients in the cities of Switzerland. It remained solidly Swiss until 1940, when it opened in New York.

First Boston, an investment bank, provides funds for economic development and global investment. Collaboration with the First Boston Corporation began in 1978 and Credit Suisse took control of First Boston in 1988. The spatial consequence was that CSFB acquired facilities in US cities and, therefore, a global presence. Today, the group has offices in 25 US cities. In 1990, CSFB bought a private bank with an offshore branch in Nassau, the Bahamas.

Other takeovers (Winterthur) enabled CSFB to enter the world of insurance in 1997, and to acquire the fourth largest Swiss bank in 1993. Other mergers took place in the USA. More recently, CSFB acquired interests in the Czech Republic, Canada and Belgium. The group now has three arms:

- CS — focused on private banking for the wealthy and commercial and retail banking in Europe
- CSFB — managing wealth and investments globally, besides carrying out market research
- Winterthur — its insurance arm, which has 69 offices in 39 countries

If office space is a guide to the relative importance of CSFB's locations, New York has 270 000 m^2 and London has 121 000 m^2 — out of a total of 604 000 m^2. There are 9000 employees in the USA, 6500 in Europe and 2400 in the Asia/Pacific region.

A sign of the group's strength in the USA is that a new global business centre opened in Raleigh-Durham, North Carolina, in 2005. In less than 30 years, CSFB has changed from a *national* bank into a truly *global* corporation.

Figure 31 shows the current global pattern of CSFB's offices.

What factors have influenced the location pattern of CSFB's offices?
- The merger of Swiss and American activities — over one-third of the offices and employment are in the USA and 10% of the offices are in Switzerland.
- A presence in all the source regions of investment funds — where there are wealthy individuals and large corporate headquarters in need of investment advice and managed investments.
- The need to have a global network of offices — which enables the group to operate 24/7. Tokyo, Hong Kong and Singapore are still open when London's working day commences. New York overlaps with London for 5 hours or more, and Tokyo just overlaps with New York but more with other US locations.
- A policy to gain presence in NICs (e.g. Kuala Lumpur, Taipei and Seoul) and RICs (e.g. Beijing, Shanghai and Mumbai) — to be able to invest and advise in these emerging regions.

Figure 31
The global spread of CSFB offices worldwide, also showing global time zones

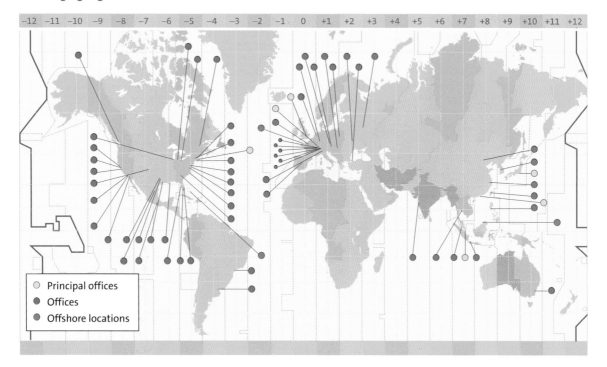

- A presence in **offshore banking** centres — to take advantage of the investment and tax regimes in offshore centres. Examples are Labuan (an island off East Malaysia), St Peter Port (on Guernsey) and Nassau (in the Bahamas). Three continents are covered by these locations.
- A presence in other major continental financial centres — for example, Sydney, São Paulo and Buenos Aires.
- Limited presence in the areas of greatest political volatility and investment risk — Africa and the Middle East.
- Some presence in the emerging states of eastern Europe — for example, Moscow, Prague and Budapest. Essentially, the branches locate where business is profitable. This is similar to other global finance corporations.

Case study 9

OFFICE LOCATION IN THE UK

The UK has a service economy dominated by London's key services — finance and business (which generated 2.6 million jobs nationally between 1981 and 2000), the public sector, tourism and hospitality, and the creative and cultural industries (Figure 32).

What determines the pattern and number of service sector jobs in the UK?
- Inertia — historically, London has been the centre of government, trading, banking and insurance. London's prestige is global and, along with Tokyo and New York, it is one of three genuinely global cities.

Figure 32
Percentage of employee jobs in service industries, London, 1997

K = Kensington
W = Westminster
I = City

Percentage of all employee jobs

- 90.0 or over
- 85.0–89.9
- 80.0–84.9
- 75.0–79.9
- less than 75

- Government policies — between 1964 and 1979, attempts were made to reduce the pressure of demand for office space in London, from both the public and private sectors. Office decentralisation policy steered firms and civil service jobs away from London (Table 14). New civil service policies in 2005, to save money, included both cuts in the number of staff and continued movement away from London.
- Planning policies in Greater London encouraged the growth of nodes in the outer suburbs — for example, Croydon and Kingston.
- Regional development policies to attract offices to development areas. Subsequent policies, such as urban regeneration and urban development corporations, have also assisted decentralisation — for example, to Canary Wharf.

Company or government body	To where?
IBM	European HQ to Portsmouth
Zurich Insurance	Portsmouth
Eagle Star	Cheltenham
Met Office	Bracknell (moved again in 2004, to Exeter)
DVLA	Swansea
Barclaycard	Northampton
Royal Mint	Llantrisant
National Savings	Durham and Glasgow

Table 14 Examples of office decentralisation in the 1970s

- Devolution of government to Scotland and Wales led to government service functions locating in Edinburgh and Cardiff.
- The need for face-to-face contact in some activities, which has only been partially replaced by ICT, with London providing the ideal location.
- The HQs of major multinational companies need to be in global centres such as London for the prestige involved.
- The growth of London's role as a global financial centre has put pressure on space in the city and forced growth elsewhere, especially in Canary Wharf.
- The growth of government research, especially in the 'golden triangle' of Heathrow, Reading and Guildford.
- The scale of most companies has necessitated the division of office work. Routine or 'back office' work, which needs no contact with the public, does not need to be located in premises with high rents. Similarly, call centre work is footloose. Face-to-face work, such as trading by investment banks and share dealing, does still depend on close contact in a small area at the heart of the city. Nevertheless, ICT has weakened these ties.
- Rental costs of premises do affect the patterns of office work. The **bid rent curve** tells us that only those businesses that can afford to pay high rents per unit area, such as finance, banking and insurance, locate at the heart of a city.
- The availability of appropriate labour with the necessary qualifications also determines office locations. For example, the law firms involved in company mergers will locate close to where company headquarters are found, and not in the suburbs.
- The break-up of the Greater London Council led to the proliferation of service jobs in all London boroughs, as did the creation of the new unitary authorities such as Blackburn.
- Decentralisation can be driven by the residential preferences of workers. Office workers about to be decentralised from London tend to have stereotypical views of industrial cities, such as Stoke-on-Trent, and are more favourably disposed to cities such as Bristol. The proposed move of BBC Sport to Manchester caused much concern among staff. On the other hand, when the Met Office moved from Bracknell to Exeter there was general accord.

(a) Identify the key factors that concentrate and disperse service jobs in London.
    Use two columns:
    ■ centripetal forces (i.e. forces pulling services into London)
    ■ centrifugal forces (i.e. those pushing services out from London)
(b) Which of the two forces will be strongest in 2050?
    Give your reasons.

The service sector in London

Finance and business services have grown more quickly in London than elsewhere in the UK — 1.5 million Londoners work in this sector, out of a total of 5.5 million nationally. Over 500 000 of these business service jobs (33% of the jobs in this sector in London) are located in the city and the eastern boroughs (mainly in Docklands). The **location quotient** (LQ) for business service jobs is 2.15. (A location quotient of 1.0 indicates that the share of jobs is at the national average.) Central London (Camden, Islington, Westminster, Kensington and Chelsea, Southwark, Lambeth and Wandsworth) has a business service LQ of 1.79 because 578 000 office jobs are located there (Table 15). Increasing numbers of employees have put pressure on office space, forcing rents up. Since 1995, business services have been expanding into west London — for example, the Paddington–Heathrow axis.

A map of location quotients extending out to the M25 shows how jobs are concentrated in certain boroughs. It also shows how the M4 corridor and the golden triangle are concentrations of financial service jobs (Figure 33).

Figure 34 covers the same area, but shows the location quotients for public service jobs, including health and education. It highlights major university and health service centres. Health and education services, such as hospitals and secondary schools, do need to be dispersed so that they are spread evenly over the population.

New types of employment in London are established in new areas, but often have a focal point of some significance to that activity:
■ Banking originally clustered close to the Bank of England.
■ The focus for insurance was the Lloyds Insurance building.
■ The media tend to focus on the West End and Mayfair.
■ The rapidly growing recruitment agencies are based in the West End, as are many ICT consultancies and support firms.

Table 15
Finance and business services employment, 2000

Region	Number of jobs	Location quotient	% change, 1995–2000
Great Britain	5 582 000	1.0	3.7
Greater London	1 514 000	1.67	5.5
East London, including city	488 000	2.15	5.6
East London, excluding city	210 000	1.31	6.8
Central London	578 000	1.79	6.1
West London	192 000	1.22	6.4
South London	168 000	1.4	6.4
North London	88 000	1.13	3.7

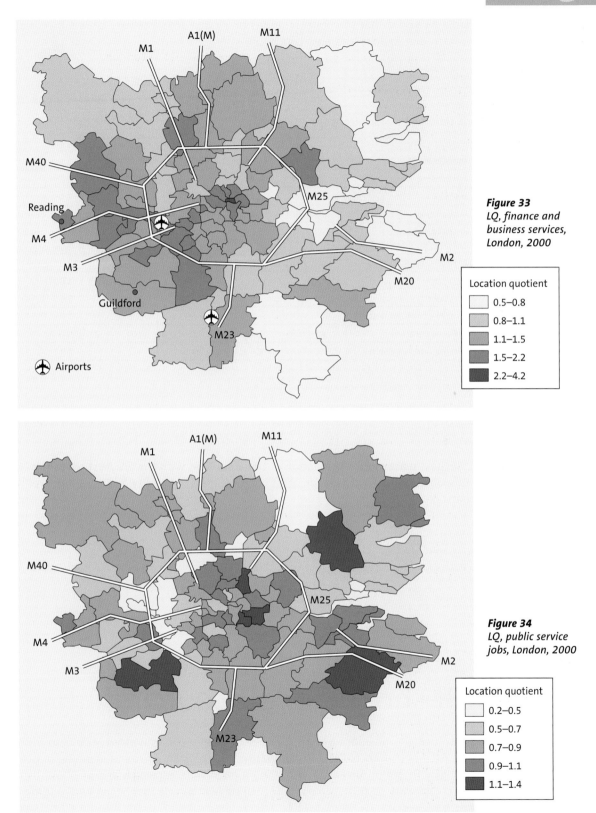

Figure 33
LQ, finance and business services, London, 2000

Location quotient
- 0.5–0.8
- 0.8–1.1
- 1.1–1.5
- 1.5–2.2
- 2.2–4.2

Figure 34
LQ, public service jobs, London, 2000

Location quotient
- 0.2–0.5
- 0.5–0.7
- 0.7–0.9
- 0.9–1.1
- 1.1–1.4

Table 16 shows that the concentrations of service activities vary between central London and the city and east London.

Table 16
Employment numbers and LQs for selected office activities

	Central London		City and east London		Greater London
	Number of jobs	LQ	Number of jobs	LQ	LQ
Banking and finance	51 200	1.5	125 000	5.1	2.1
Broking and fund management	20 900	3.2	41 000	9.0	3.6
Advertising	28 900	4.6	ND	ND	2.5
Legal	35 500	2.2	46 200	4.0	2.1
Accounting	38 900	2.8	23 600	1.4	2.0
Recruitment and business support	147 200	1.9	92 500	1.7	1.7
Market research	50 100	3.0	13 200	1.1	2.0
Insurance	ND	ND	22 000	2.2	1.0
Computing	50 100	1.5	25 700	1.1	1.6
Property development	36 300	2.4	17 500	1.6	1.7
Higher education	43 000	1.54	ND	ND	ND
Cinema and television	65 300	4.6	12 400	1.2	2.9

Concentration in the city and east London
Concentration in central London

Question

Study Table 16. Suggest reasons for the location of those services with the greatest concentration in central London and those with the greatest concentration in the city and east London (including Docklands).

Guidance

Remember: any location quotient above 1 is above average concentration.
Think in terms of services that need to be localised, those that agglomerate and those that need a central location.

Global shift of services: off-shoring or outsourcing

Outsourcing or off-shoring is the global shift of services away from MEDCs to NICs and RICs and even LEDCs. In comparison to the shift of manufacturing, it is a relatively recent occurrence (post-1995). Global shift, especially the off-shoring of call centre work, is seen as a possible final stage in a cycle that began two centuries ago (see Part 3):

- Stage 1: Collapse of the Indian textile industry in the nineteenth century

The starting point for the cycle was arguably the impact of British colonial rule on Indian cotton textiles. **Protectionism** was a common colonial policy, removing competition and enhancing export markets for manufactured goods. Interestingly, the British also insisted on Indians learning English in schools.

- Stage 2: Internationalisation, 1920–39 and 1950s

Internationalisation and the development of what were to become transnational companies (TNCs) began in the automobile industry. Both Ford and GM opened assembly plants in Europe — for example, GM took over Opel, at Russelsheim near Frankfurt. Chemical and electrical engineering companies also began to open plants on either side of the Atlantic.

- Stage 3: Rise of new industrial producers and deindustrialisation, 1960s

A foretaste of what was to affect the service sector came in the 1960s, when the Japanese and South Korean governments backed the development of heavy industry, especially ship building and steel making. Deindustrialisation of areas such as the northeast of England, the Clyde, Birkenhead and the Saarland started. Over the next 10 years, steel production in the UK began its long process of rationalisation. Textiles and clothing manufacture also began to shift to south Asia.

- Stage 4: Global shift and branch plant economies, 1980s

The global shift of major industries took off. Some companies had internationalised earlier — for example, VW in Brazil and Mexico. Japanese car assembly plants were established in the USA and the EU — for example, Toyota and Nissan. The first shift of car assembly from MEDCs to NICs occurred — for example, Proton at Shah Alam, Malaysia, assembling Mitsubishi designs.

Electronics branch plant assembly lines in NICs and RICs were established — for example, Bosch in Penang, Malaysia. As labour costs rose, many of the plants in South Korea and Singapore were moved to lower labour cost locations in Indonesia, Vietnam and China. Firms established in NICs were also opening branch plants in neighbouring NICs and RICs — for example, Advent moved from Taiwan to Malaysia. Designer labels outsourced to NICs and to 'Arab' world countries — for example, CK jeans are made in Tunisia and Hugo Boss clothing is made in Turkey.

A network of regional financial service centres grew to supplement the activities in the primary locations (London, New York and Tokyo). Hong Kong, Singapore, Sydney, Dubai and Mumbai grew as secondary financial centres. These, together with European centres such as Paris, Frankfurt and Milan, provided conditions that would enable the financial services sector to straddle the world. Outsourcing of call centre work started in 1981, when Infosys opened in India.

- Stage 5a: Production in the former Soviet bloc, post-1989

After 1989, eastern Europe opened up to industrial investment, primarily from the EU. Some companies, such as Polski Fiat and VW/Skoda, had already invested in the region. This confirms the trend, noted by Paul Dicken in his book *Global Shift* (1998), for companies to focus a large part of their investments within their global region rather than globally.

- Stage 5b: Rise of call centre outsourcing, 1990–2000

  After 1991, the shift of call centres to south Asia took off. China and India joined the world economy as more companies opened plants or made agreements to assemble products.

- Stage 6: Business process outsourcing or off-shoring, 2000 onwards

  There are two types of off-shoring: back office or business process outsourcing, such as accounting, and customer-facing activities, such as call centres.

In 2003, approximately 8000 customer-facing jobs out of 800 000 were transferred from the UK to India. There were still 56 000 call centre jobs in Scotland. Nevertheless, India and the Indian Ocean rim is forecast to receive 2 million jobs from MEDCs by 2008. The Indian call centre work grew by 60% between 2000 and 2003. It is estimated that 2 million of the 13 million jobs in financial services in MEDC economies will have moved to India by 2008, 730 000 of which will be jobs originating from Europe.

There are two types of outsource working:
- setting up an overseas offshoot, known as a captive
- buying in the service from other companies

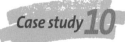

Case study 10   **WHO IS GOING WHERE?**

Why outsource to India?
Table 17 shows which companies have outsourced service sector jobs to India.

There are several reasons why India is a common destination for outsourcing:
- the Commonwealth link with the UK
- English language skills are good and most outsourcing comes from the USA and the UK
- strength of IT in the education system in India
- low communication costs — sending car hire calls to Bangalore rather than a UK centre costs 20p for the whole transaction, which is a cost reduction of up to 50%
- wages are lower — in 2004, a US IT manager cost $55 000, whereas the cost in India was $8500
- India produces 3 million graduates a year
- the multiplier effect of investments in IT in Bangalore and other cities, such as Chennai and Hyderabad (which are replacing Bangalore as the prime destination); half of the world's top 500 companies now outsource IT or other business processes to India
- Indian specialist off-shoring companies are growing — for example, ICICI has four centres in Bangalore and one opening in Mumbai in 2004
- capital costs are lower

Indian call centres: a staff shortage
The last thing you would expect India's call centre bosses to be worrying about is a shortage of staff. The 'business process outsourcing' (**BPO**) industry employs an estimated 348 000 people. Nearly 3 million English speakers graduate from Indian

Banks	Original location	Details
Abbey National	UK	100 skilled IT staff to Hyderabad
ABN Amro	The Netherlands	2000 staff at call centres handling credit risk and payment authorisation in Chennai
Amex	USA	4000 financial accounting and data management staff — a captive company
Aviva (Norwich Union)	UK	3700 jobs in call and claims processing — already 1200 posts outsourced to New Delhi
AXA	UK	100 back office posts
Barclays Bank	UK	150 business banking and credit card jobs (2003)
Barclays Capital	UK	400 to Singapore (2004)
British Airways	UK	2400 posts, managing passenger accounting, error handling and frequent flier miles (1996), opened in Mumbai — a captive company
BT	UK	Plans for 2200 jobs and possibly up to 7000 with HCL technologies in New Delhi
Capital One	USA	1200 posts in customer services, risk and production services
Citibank	USA	3000 posts in wholesale and retail banking
CSFB	Swiss/USA	Online retail banking with Cognizant
Delta Airlines	USA	Booking centres in Mumbai and Philippines
Deutsche Network Services	Germany	50 payment and management processing jobs
Fidelity Investments	UK	200–250 posts initially, rising to 1000 posts in a business process outsourcing centre
Goldman Sachs	USA	250 jobs created
IBM	USA	1000 jobs planned to move, 2004 onwards
HSBC	UK	4000 posts gone from Swansea, Birmingham, Sheffield and Brentwood; currently, 2600 to Hyderabad, 1900 to Bangalore, 2800 to Guangzhou and Shanghai, China, and 600 to Kuala Lumpur, Malaysia — mainly accounting and transaction processing — a captive company
JP Morgan Chase	USA	3000 employees in transaction processing
KeyBank	USA	IT support with WIPRO at Electronic City, Bangalore
Lehman Brothers	USA	Outsourcing IT services to WIPRO, Infosys and Tata
Lloyds TSB	UK	Closing Newcastle call centre and outsourcing to Hyderabad
Morgan Stanley	USA	1600 jobs to provide transaction support
National Rail Enquiries	UK	600 call centre jobs
Prudential	UK	Plans to employ 250 people in Mumbai by 2004, rising to 850 — will save $25 million per annum
Standard Chartered Bank	USA	4500 employees

Table 17 Outsourced service sector jobs, India

universities every year. This has been India's big attraction: call centres have been spoilt for choice. Finding and retaining qualified workers has become the industry's biggest worry and it may stall the offshore call centre boom.

The industry has grown explosively. Youngsters have hopped from job to job. Staff attrition rates for the industry have climbed to 45–50% a year. Starting salaries have reached about 10 000 rupees a month ($230), considered very high for a first job (which also explains why outsourcing to India remains so attractive). Training costs are also mounting as firms take on less-qualified applicants. The president of the Call Centre Association of India (CCAI), which represents some 60 of India's 400 BPO firms, says the pressure is such that firms do not always check staff references. This has led to a few, widely publicised, fraud scandals.

With such fast growth, training has become much less organised. The industry is beginning to help itself. The CCAI, with the Confederation of Indian Industry, has launched a training initiative. It will offer a standardised qualification for new BPO workers — improving English, 'neutralising' accents and teaching some computer skills. NASSCOM, a lobby for the software and services industry, is also introducing an 'assessment and certification' programme for would-be employees. Such schemes should cut costs and ease wage pressures.

India still has a unique combination of workers and English-language skills, but the full potential of BPO, beyond call centres, has only been glimpsed — there are huge emerging markets in legal services, accounting, healthcare and personnel services. It would be a shame if India were to miss out by misusing its unbeatable, seemingly inexhaustible resource: well-educated young people.

The Philippines

Table 18 gives six examples of outsourcing to the Philippines. It is noteworthy that the USA dominates because of its former colonial links.

Table 18 Examples of services outsourced to the Philippines

Company	Original location	Details
Alitalia	Italy	Accounting
AOL	USA	Billing
Arthur Andersen	USA	Software production
Barnes & Noble	USA	Online purchasing
Mitsubishi	Japan	Engineering design
The Red Cross	Switzerland	Accounting

The global picture

China, Malaysia, Singapore, Australia, South Africa and Vietnam are all destinations for off-shoring. Now countries such as Ghana, Barbados and Bangladesh are being considered. Your GCSE examination answers may have been marked online in Australia!

There are regionalised job flows:

- from the UK to countries of the Commonwealth and especially south and southeast Asia and South Africa
- from the USA to Latin America, and to south and southeast Asia (e.g. the Philippines)
- from Japan to east and southeast Asia
- from France to former French colonies
- from the EU to eastern Europe

Outsourcing within the 'New Europe'

'New Europe' covers the east European countries that have joined, or are candidates for joining, the EU. The ten countries that joined the EU in 2004 opened up new locations for outsourcing. The reasons are:

- these low-labour cost locations are closer than Asia
- they are within the EU trading bloc
- the linguistic skills of many of the new EU members extend beyond English to include other major European languages, especially German

Manufacturing has also moved east (see Part 4). Recently, EU business services have moved east in response to the cheaper costs obtained by UK and US companies in Asia. In terms of attractiveness to businesses, the Czech Republic, Poland and Hungary are highly rated by companies as locations for outsourcing (Table 19).

Country	Town/city	Company	Original location	Examples of activities
Poland	Warsaw	IBM	USA	IT services, human resources, accounting, payroll, taxes, financial services, back office functions, transaction handling
	Krakow	KPMG	UK	
		Lufthansa	Germany	
		Capgemini	France	
		IBM	USA	
	Lodz	Philips	The Netherlands	
	Poznan	KPMG	UK	
	Bielso-Biala	Fiat	Italy	
	Olsztyn	Citibank	USA	
Czech Republic	Prague	Accenture	UK	Financial management, customer support, IT services and logistics, parcel tracking, business services
		DHL	USA	
		Hewlett-Packard	USA	
		ExxonMobil	USA	
		IBM	USA	
		Siemens	Germany	
		Tesco	UK	
Hungary	Budapest	GFT	UK	Services for banks, finance centre IT support and accounting, European service centres, logistics, training
		Diageo	UK	
		General Electric	USA	
		ALCOA	USA	
	Szekesfehervar	Lidl	Germany	
		Oracle	USA	
Slovakia	Bratislava	Dell	USA	Customer support, financial services, call centres
		IBM	USA	
		BA	UK	
		Hewlett-Packard	USA	

Table 19 Examples of outsourcing to eastern Europe

Proximity is cited as important by some companies and, therefore, eastern Europe is particularly attractive to German companies. Outsourcing has led to 3000 jobs moving to Poland, a total that could rise to 200 000 by 2008 as European firms restructure. As well as European firms, several US multinational companies with a strong European base are moving activities — examples are IBM to Krakow, Warsaw, Prague and Bratislava; Capgemini Accounting to Krakow, and Dell and Hewlett-Packard to Bratislava. The fact that some new destinations are in the same time zone as Germany and France and only 1 hour ahead of the UK is a definite advantage over India (which is 5 hours ahead of GMT).

Question

Look at the data in Figure 35.
Make up a ranking table to show the top five countries in terms of
- financial structure, i.e. costs
- business environment
- people skills and availability

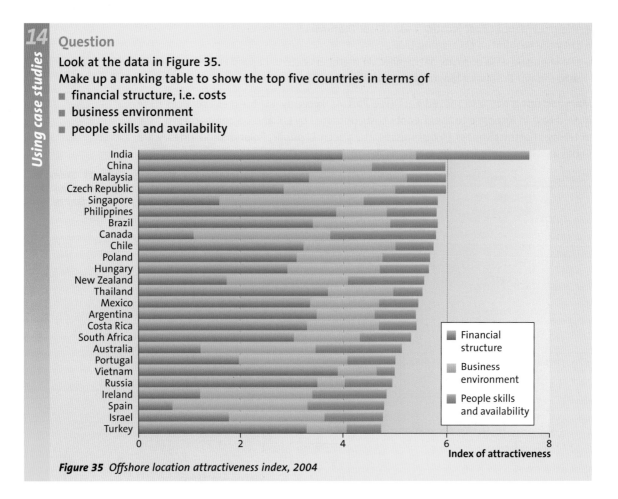

Figure 35 *Offshore location attractiveness index, 2004*

Calculating the impacts

What are the costs for RICs such as India?

- Westernisation and loss of cultural identity as a result of training and conversing with English speakers — 'abandoning identity and slipping into someone else's' (*Monbiot*).
- Abuse from angry customers might result in anti-Western attitudes.
- Some employers call their programmers 'slave coders' — hardly a term to give the workers a feeling of worth.
- Unsocial hours — the time difference between India and the UK (5 hours) and the USA (10–13 hours) makes night working essential.
- High staff turnover in call centres — with most employees leaving after 3 years without skills that can be transferred to other jobs.
- Increasing social divisions between those who have the new jobs and those who do not. 'You can't have a scenario where you have a small part of a nation's population entering an advanced league while everyone else stays behind. If you

have an island of affluence in a sea of poverty it is not sustainable' (*Nandan Nilekani, Infosys*).

■ Further relocation will occur, as happened when labour costs rose in Malaysia and Singapore. Already, Bangalore's call centre operations face rising costs as employees move to higher salaries.

What are the benefits for India?

■ An estimated $24 billion benefit to the Indian economy by 2008.
■ Call centres and back office services are forecast to employ 3 million people by 2005 — but note that this is only 0.3% of the total population.
■ More jobs are created in the receiving country than are lost in the MEDCs.
■ Young workers have a higher disposable income. Most of these are graduates with an average age of 23. Infosys received 600 000 applications for 7000 jobs in 2002, indicating the demand for this work. 80% of employees are 20–25 years old.
■ Gender apartheid is being reduced because the call centre jobs are split 50:50 male:female. (In the UK, the ratio is 20% male to 80% female.)
■ Employment with higher starting salaries.

What are the costs for MEDCs?

■ UK trade unions estimate that there will be up to 200 000 job losses in financial services in the UK by 2008, and 500 000 are expected in the USA. By 2003, 50 000 jobs had already gone to cheaper markets. It has been described as 'the hollowing out of IT' in search of lowest costs. It is estimated that 2 million of the 13 million jobs in financial services in MEDCs will move by 2008.
■ There is a loss of female jobs. The work is currently 80% female in the UK.
■ Many outsourced jobs were established in vulnerable deindustrialised areas.
■ There can be a legislative backlash. In 2003, New Jersey and Indiana banned outsourcing to Indian companies.
■ Skilled service activities are also moving. For example, medical image analysis has moved from the USA and Barclays Capital moved 400 jobs from Canary Wharf to Singapore in 2004.
■ 'When you have a software developer having to train his replacement in India, you realise this is not about skills. This is about a global economy that is based on the lowest-cost labour, and multinationals exploit that' (*The Washington Alliance*).
■ Some companies, such as Lehman Brothers and Dell, are taking work back because of a lack of quality. This is a small, costly, reverse flow.

What are the benefits for MEDCs?

■ Many companies cite efficiency gains — 'We are moving jobs from one part of the company to another' (*HSBC*).
■ Improved profits benefit the shareholders, such as those with pension funds. It has been estimated that if £1 is spent in India, it creates £1.47, of which £1.14 comes back to the UK as profit.
■ The world's top 100 financial institutions will have saved $138 billion by 2005, with an average 39% savings in costs.

- The labour costs of a call centre worker in the UK are £18 000, compared with £2500 in India.
- There is a desire to 'follow the crowd'. 'Our competitors have got this religion' (*Microsoft*). 'One of our challenges is to balance what business needs to do with the impact on people and this is one of those areas where this challenge hits us straight between the eyes. Our competitors are doing it, so we have to do it.' (*IBM*).
- Outsourcing enables firms to focus on new and innovative products in the country that is outsourcing, rather than focus on older products.

Where will it all end?

As well as outsourcing services and office functions in a variety of skilled professions, global companies are beginning to outsource their research and development, which was traditionally done at headquarters. A range of health, education, planning and architectural functions are being outsourced to a wide variety of countries.

15 *Using case studies*

Question

With reference to a range of examples, write a balanced assessment of the costs and benefits of outsourcing services.

Guidance

- Always support your assertions with detailed examples.
- You need to look at a number of players — companies, workers and consumers — and also at MEDCs, NICs/RICs and LEDCs.

Making the world more equal

The biggest challenge of globalisation is to spread improvements in the quality of life to all. The world contains 200 grossly unequal countries, which are becoming more unequal (see Part 2 and Table 20).

The number of people worldwide living in extreme poverty is falling slowly — from 1451 million in 1981 to 1101 million in 2001. In Asia, the number fell from 606 million in 1981 to 212 million in 2001. Between 1990 and 2001, 30 LEDCs experienced greater than 3 % economic growth and a further 71 countries managed some growth. However, there are still 54 countries, with a total population of 750 million people, where incomes per capita are declining. These countries are being left behind — lacking inward investment, but with disease, poverty and civil disorder. In 2002, Singapore received more inward foreign investment than all of sub-Saharan Africa. In this part, we explore the roles of trade, investment and aid in bridging the development gap.

Year	Ratio
1820	1:5
1913	1:13
1950	1:33
2004	1:100

Table 20 *Ratio of wealth between the richest and poorest countries, 1820–2004*

Trade

Trade is one way to enhance growth and development and to eliminate poverty. This might be the case in theory, but the 49 LDCs that make up the world's poorest countries have not shared in the growth of world trade. The same number of people (approximately 645 million) live in the top five exporting countries (the USA, Germany, Japan, France and the UK) and they have 100 times more trade than their poor counterparts. About 70 % of the world's poor live in rural areas and work in agriculture; yet, because of subsidies amounting to $330 billion, 66% of agricultural trade is exports from the OECD (Organisation for Economic Cooperation and Development) countries. 'We are told that trade can provide a ladder to a better life and deliver us from poverty and despair. Sadly, the reality of the international trading system today does not match the rhetoric' (*Kofi Annan, Secretary General of the UN, 2003*).

One of the foundations of the current patterns of trade is the impact of history. London was the centre of trading innovation in the eighteenth century, which stimulated trading, led to the Industrial Revolution and was the basis of London's current financial status. The trading advantages that Europe and the USA have held for over a century have increased, and only Japan has gained a similar advantage in the twentieth century. In the past 25 years, there has been a slight shift towards Asia, with the rise of the NICs and RICs (see Part 4).

Meeting the challenge in the last 50 years

The International Monetary Fund (IMF), formed in 1945, is an organisation of 184 countries, working to foster global monetary cooperation, secure financial stability, facilitate international trade, promote high employment and sustainable economic growth, and reduce poverty. The USA has a veto on decisions and holds 17% of the votes on any decision. The IMF's activities are almost certain to be influenced by the 'big five', as they have 39% of votes, or by the MEDCs in general, as they have 66% of votes. Some argue, therefore, that the IMF works in the interests of MEDCs by imposing conditions on loans that reflect the ethos of the developed, capitalist world.

The World Bank, formed in 1944, focuses on reconstruction after natural disasters (e.g. the Asian tsunami), on humanitarian emergencies (e.g. the Sahel famine) and on post-conflict rehabilitation needs that affect developing economies (e.g. Bosnia). Recently, the World Bank has focused on poverty reduction. It has not been without criticism — for instance, for supporting the multinational pharmaceutical companies in their health policy work for the African AIDS pandemic.

The World Trade Organization (WTO), formed in 1995, replaced the General Agreement on Tariffs and Trade (GATT). GATT was formed in 1947, and the WTO continues GATT's process of trade liberalisation, obtaining agreement to remove tariffs on products. The WTO is seen by many as an organisation that favours MEDCs. Its Seattle Conference (1999) was a focus for many anti-globalisation groups, besides more formal protests from LEDCs. The Doha Conference (2001) focused on trade in cash crops, such as cotton, sugar and cocoa, and not least the dumping of products such as EU sugar. The conference in Cancun (2003) made little progress because LEDCs were concerned by:
- access to MEDC markets
- protectionism by industrialised countries and regions such as the USA and the EU
- rich countries dumping agricultural commodities on international markets at prices below the cost of production
- patent rules used by pharmaceutical companies that appear to deny LDCs access to affordable medicines

MEDCs, especially the G8, dominate 75% of global trade and only 17% of trade is from the high income countries to LEDCs. Goods movement between developing countries forms only 6% of global trade. That is not to say that trade does not improve the development prospects of some countries. In 2002, 68% of the imports of OECD countries from middle-income countries were of manufactured goods; this represented a 46% rise since 1992 and reflected the growth of NICs and RICs. Some progress has been made in LDCs, so that more manufactured goods are being exported.

16

Question

Figures 36 and 37 compare the trade flows to and from Africa, which has relatively little industry, and the trade flows to and from Europe. Write a detailed comparative analysis of the pattern of trade flows shown in Figures 36 and 37.

Guidance

■ Always use precise data — i.e. measure the width of the flow lines to assess volume, and note specific inwards and outwards flows.

■ Comparative statements are required about total volume and particular links.

■ In an analysis, you need to try to suggest some reasons.

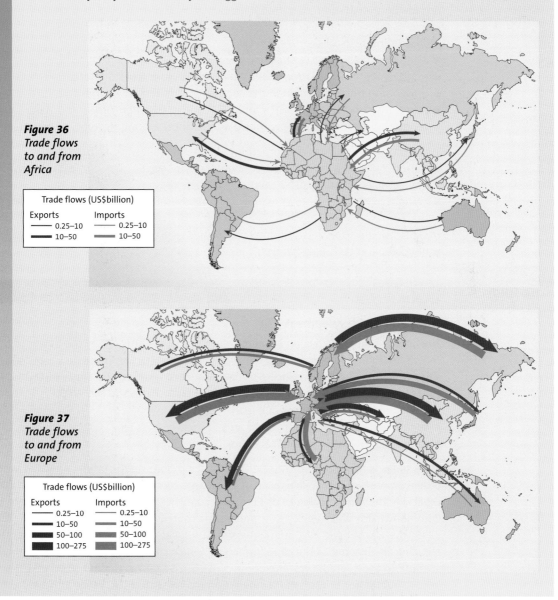

**Figure 36
Trade flows
to and from
Africa**

Trade flows (US$billion)

Exports	Imports
—— 0.25–10	—— 0.25–10
—— 10–50	—— 10–50

**Figure 37
Trade flows
to and from
Europe**

Trade flows (US$billion)

Exports	Imports
—— 0.25–10	—— 0.25–10
—— 10–50	—— 10–50
—— 50–100	—— 50–100
—— 100–275	—— 100–275

Some of the continuing barriers to trade that LDCs and LEDCs face are:

- protection of agriculture by MEDCs, so much so that agricultural exports to LEDCs are considerably more valuable than manufacturing exports
- tariffs on industrial products being higher in LEDCs — relaxing these would harm LEDC manufacturing
- trade blocs (see below) having high tariffs — for example, industrial products from Latin America are faced by tariffs in Africa that are six times higher than those for goods from Europe

The role of trade blocs

A trade bloc is an intergovernmental association that manages and promotes trade activities for a specific global region. Trade blocs can threaten free trade and protect the trade interests of their region by:

- establishing some control over trade that meets the interests of members
- establishing tariffs to protect intra-bloc trade from 'outside' forces
- developing trade to enhance the security in the region
- promoting trade among LDCs and LEDCs (e.g. between Africa and Asia)
- promoting economic and technical cooperation among developing countries

Trade blocs try to restrain competition, using the following tactics:

- import quotas — limiting the amount of imports, so that domestic consumers buy products made by the countries in their region
- being bureaucratic — to slow down the ability of imports to enter the domestic market
- subsidising sectors of the home economies — to gain competitive advantage
- boycotts and technical barriers

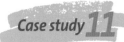

Case study 11 | SUGAR

Trade and the development gap

The impact that trade has on economic development is affected by the power of both the multinational companies and the actions of the WTO. This is the case especially for small-scale agricultural producers of bananas, cocoa, coffee and sugar in LEDCs. The sugar trade illustrates the effects of policies of trade blocs and the WTO.

There are two main sources of sugar:

- sugar beet — grown in temperate areas
- cane sugar — grown in tropical areas

Cane sugar is adaptable to tropical climatic conditions and is often cultivated where other commercial crops have failed. It is a staple commercial product of many LEDCs, especially the former colonies of European countries. In the UK, cane sugar from the Commonwealth was subject to a preferential purchase agreement after 1928. After 1945, the UK bought all Commonwealth surpluses at below the world price, thus reducing the income of the cane farmers — which can be interpreted as a colonialist attitude to those producers. The Commonwealth Sugar Agreement (CSA) of 1951 set import quotas for Britain and a single Commonwealth price. While this held prices down in post-war UK, it did little to give cane producers a viable income.

Europe and sugar

In 1958, the EEC granted financial aid to the former European colonies in Caribbean, African and Pacific (CAP) countries. Most of these countries gained their independence and signed the Yaoundé Convention of 1963. This convention initiated trade and aid agreements between an enlarging EEC and the CAP states, a group of 77 developing countries from Africa (48), the Pacific (14) and the Caribbean (15).

Enlargement of the EEC, with the accession of the UK, Ireland and Denmark in 1973, led to Commonwealth countries in Africa, the Caribbean and the Pacific seeking to establish cooperation with the enlarged EEC. The UK was anxious to support the Commonwealth sugar producers. The EEC and the CAP states signed the Lomé Convention (1975), which changed the UK commitment to the Commonwealth (excluding Australia) into an EU commitment to the CAP states. Subsequently, the EU and the CAP states signed the Cotonou agreement (2000), which included a 'sugar protocol' that maintained a long-standing UK commitment to maintain preferential access for CAP sugar.

The economies of the CAP states and 48 LDCs are dependent on the preferential agreements in place for sugar — in particular, guarantees of stable prices, which provide a sound basis for long-term investment in those countries. The socioeconomic importance of sugar in CAP states is such that the erosion of their preferential access would result in extremely negative consequences for employment (300 000 jobs currently), social services, GDP, exports and energy production. The CAP agreement ends in 2008.

The EU is the world's second largest importer of sugar. Two-thirds of this (about 1.1 million tonnes) is imported into the UK from the CAP states and LDCs.

Current sugar production

Global sugar production barely meets demands (see Table 21). Output in LEDCs is forecast to reach 99.2 million tonnes — reflecting continued growth in Brazil (27 million tonnes), where it is used as a petrol substitute and for producing alcohol as well as sugar, and Africa.

- Brazil's output is rising, aided by price reductions caused by its currency's devaluation. Sugar sales are subsidised by the bioethanol sales for powering cars.
- Hurricanes have affected production in the Caribbean (there was a 15% reduction in Jamaica).
- Production in the Far East is 41 million tonnes.
- Output in India is rising, although the 2004 tsunami ruined the crop in Tamil Nadu.
- Drought has affected output in China.
- Production in Africa is 7.9 million tonnes, of which 2.6 million tonnes are from South Africa.
- Production in the USA is 8 million tonnes, but this has been affected by hurricanes that hit cane growers in Florida and Louisiana.

	Production, 2004 (million tonnes)	Consumption, 2005 (million tonnes)
Latin America and the Caribbean	47.8	26.1
Africa	5.3	8.1
Near East	5.7	11.0
Far East	42.1	50.9
Oceania	5.6	1.5
Europe	21.8	20.3
of which EU	17.8	14.9
North America	8.2	10.9
Former Soviet countries	4.0	11.7
Others	3.5	4.3
of which South Africa	2.6	1.6
Total world	144.0	144.8

Table 21 *Global production and consumption of sugar, 2004–05*

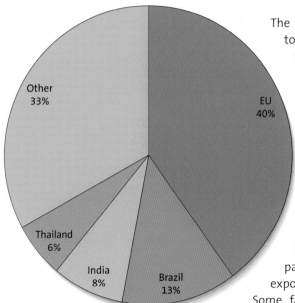

Figure 38 *World white sugar exports: percentage share of the world market, 2000–01*

The production of beet sugar in MEDCs is 42.6 million tonnes. This is mainly attributed to a 5.7% increase in the EU output. Output is rising in France (4.6 million tonnes) and Germany (4.2 million tonnes). Global sugar consumption was estimated by the FAO to be 143.2 million tonnes in 2004, increasing to 144.8 million tonnes in 2005 (Table 21). Developing countries account for most of the increased consumption.

Sugar trade

Figure 38 shows that 40% of the world sugar trade originates in the EU. Why is this the case? The EU has set production quotas for which an intervention price (three times the world price) is paid. Some of the quotas (3.1 million tonnes) were exported with an export refund — effectively a subsidy. Some farmers in the UK, France and Germany grew an excess of beet (3.8 million tonnes) and this was exported or dumped on the world market without a subsidy. In addition, tariffs — officially called 'border protection' — of 140% protected European sugar producers. Overall, European sugar production was subsidised and the EU subsidised the dumping of sugar on world markets, with an estimated cost to European taxpayers of €1.6 billion. A similar policy operates in the USA, where high prices, low cost loans to producers and a 150% tariff have supported domestic production. Japan also has a protectionist policy. It is estimated by the World Bank that up to 80% of world sugar production is traded at subsidised or protected prices.

Who gains?
- sugar producers in the EU with four companies controlling 50% of production
- sugar farmers, especially the large-scale producers
- the 17 CAP states who are part of the Sugar Protocol, which is seen as a form of aid
- the efficient exporters — Brazil, Thailand and Australia

Who loses?
- the environment, because of the heavy use of fertilisers and pesticides
- the 60 CAP states that are not part of the Sugar Protocol
- the CAP states that try to process sugar and add value when prices are forced down by subsidised dumping
- other producers in low-income countries where export prices have been forced down, e.g. India and Cuba; Cuba has had to resort to organic farming as it suffers from a US trade embargo and lost its protected access to USSR markets
- African producers who cannot compete with EU dumping, for example in north Africa

WTO intervention

In 2004, the WTO found against the EU's practice of dumping sugar. The WTO ruled that about half of the EU's annual 5 million tonnes of sugar exports breached world trade rules. The EU had to face up to the reality that it was twisting the rules to favour its own industries at the expense of the environment at home and producers in LEDCs.

Contemporary Case Studies

Importantly, the ruling did not affect the right of the EU to import sugar from the CAP and India on preferential terms.

European Commission's proposals on sugar sector reform

Subsequently, the European Commission made proposals to reform the sugar sector. The key proposals are:

- a 37% cut in support prices for sugar beet and raw cane sugar over 3 years
- a reduction in EU production quotas over 4 years to take 2.8 million tonnes (16%) out of production
- compensation for EU beet growers for 60% of their income loss over 3 years
- imports from the LDCs to be retained under the 'Everything But Arms' agreement, 2001
- a review by the European Commission of price and quota levels in 2008

The programme proposed by the European Commission implies the following:

- There will be a substantial reduction in sugar beet production in the EU and the concentration of production in the most efficient producing areas. EU production will shrink to between 8 million and 14.5 million tonnes. There is a risk of Europe having a sugar supply deficit.
- A significant reduction in CAP and LDC income as cane suppliers to the EU receive lower prices and most developing country producers struggle to compete.
- The introduction of greater flexibility and competition for beet and cane sugar supplies within the remaining EU processing and refining sectors.
- Commercial sugar-using industries will receive lower prices, although there is no guarantee that lower prices will reach the shops.

The effect on Ireland, an EU member

The Irish sugar beet industry is worth €140 million in total. The government, Irish Sugar (the refiner), beet growers, employees, road haulage contractors and the communities in Mallow and Carlow, where Ireland's two processing plants are based, are concerned about the threat to their sugar industry.

The reform of the sugar regime was unavoidable due to developments at the WTO. There has been a backlash to the European Commission proposals to abolish intervention, substantially cut back export refunds, lower the sugar price and reduce overall EU sugar production between 2005 and 2008. A proposed 37% cut in the beet price by 2007 and a European quota reduction of 16% by 2008 were described as unacceptable by Ireland. Europe's 332 000 beet growers, including 3800 in Ireland, are opposed to the overhaul of Europe's €5.3 billion sugar market.

The proposals are a serious threat to the Irish sugar beet industry, worth €75 million annually to the beet growers, and could result in the closure of one of the two processing factories. The factories employ 650 workers and are the economic base of both Mallow and Carlow. Hauliers transport 1.3 million tonnes of beet to the plants each year to manufacture the Irish quota of 199 208 tonnes of sugar. Up to 8000 jobs depend directly and indirectly on sugar production in Ireland, when upstream and downstream supply and service industries and seasonal employment (employment multiplier) are included. Beet growing in Ireland might become totally uneconomic. The European Commission insists the current system of sugar production is not sustainable in the long run.

(Based on an article in *Irish Examiner*, 8 January 2005)

The effect of an open market on Barbados — a CAP state

The cane sugar industry in Barbados faces an uncertain future. Much depends on Barbados finding other profitable uses for sugar cane, such as the generation of electricity and using it as a motor vehicle fuel (bioethanol). Barbadians fear that they will not be able to compete. Barbados can import sugar much more cheaply than it can produce it. If Barbados is to remain a sugar-cane-growing country, but not necessarily an important sugar producer, it will have to follow the example set by Brazil and use sugar cane to produce ethanol for cars.

The EU has assured Caricom sugar producers — Barbados, Belize, Guyana, Jamaica, Trinidad and Tobago and St Kitts-Nevis — as well as the APC states, that the EU would continue to provide them with preferential access to its markets. Caricom produces 700 000 tons of raw sugar every year, half of which is exported to Europe.

The WTO ruled in 2004 that Europe's practice of subsidising its sugar producers was a flagrant violation of international trade rules. The EU has already announced plans to reduce the price it pays to APC sugar producers. But instead of cutting the prices by 3%, the EU has decided to delay the price reduction until 2006. Caribbean producers of cane sugar are achieving relatively low yields per hectare of land compared with Australia, Brazil and India, and it is these who would stand to benefit more from an expansion in world sugar trade.

(Based on an article in *Nation News*, Barbados, 9 January 2005)

The story of sugar illustrates the ways in which MEDCs (the EU, the USA and Japan) have manipulated trade rules to their advantage. It also shows that LEDCs can get the WTO to halt such practices in the interests of fair trade. Changing the production rules can have a disastrous effect on the economies of many countries, especially when adjusting the beet/cane balance.

The impact of unfair trade will take time to heal. However, the rise of artificial sweeteners and a health-conscious developed world might take away some of the advantages that have been gained for cane sugar producers. This case study represents a single example of the complexities of trading among the world's nations when there are so many vested interests.

Fair trade

If free trade has its problems in increasing inequalities, i.e. it fails the poor, what is the future of **fair trade** as a means of helping LDCs develop?

Fair trade can be defined as a trading relationship whereby buyers in rich MEDCs guarantee a fair price with minimal fluctuations to producers of goods in developing countries. According to the Fair Trade Foundation, equitable trade (i.e. fair trade) will ensure that developing world producers have a just return for their work and an improved quality of life.

The fair trade movement has grown dramatically in recent years and now operates in 20 MEDCs, which buy produce from farmers and cooperatives in 50 developing countries. The concept of fair trade runs parallel to the need to modify trade rules and liberalise trade in a way that is fairer to the world's poorest countries. It is

Fair trade products are becoming a common sight in our shops

fuelled by the purchasing power of educated customers who will pay a slightly higher price for a more ethical transaction. Fair trade is a niche market, but many of its advocates see it as better than just giving to charity, and more sustainable because it is a two-way exchange.

It is estimated that over 5 million of the world's poorest people are experiencing tangible benefits, such as health and education improvements, from fair trade in textiles, cocoa, chocolate and coffee. At the 2005 Gleneagles summit, G8 participants began to talk about the contribution of fair trade towards 'making poverty history'. There are concerns that even fair trade could become over-commercialised as big organisations adopt fair trade suppliers. In addition, as with any other trade, over-production could become a problem when over-dependency sets in. A recent development involves part ownership of some of the commercial outlets, such as coffee shops, by the LDCs cooperatives. Fair trade products in the UK had a value of £195 million in 2005, compared with £16.7 million in 1989.

> Marks and Spencer announced it would henceforth only sell tea and coffee from certified ethical developing world sources. Topshop is also in on the act, with a trial of fair trade clothing, as are the Sainsbury's group and Virgin. They have worked out, there's money in goods that don't screw the little guy. But when we buy these products, are we really putting extra pennies in the pocket of the Ghanaian coffee farmer, or are we just appeasing the economic-imperialist guilt?
>
> The *Guardian*, 8 March 2006

Foreign direct investment

Foreign direct investment (FDI) is a direct investment that occurs across national boundaries when a TNC or global company sets up a branch or invests in a firm in another country. FDI is a measure of foreign ownership of factories, mines and

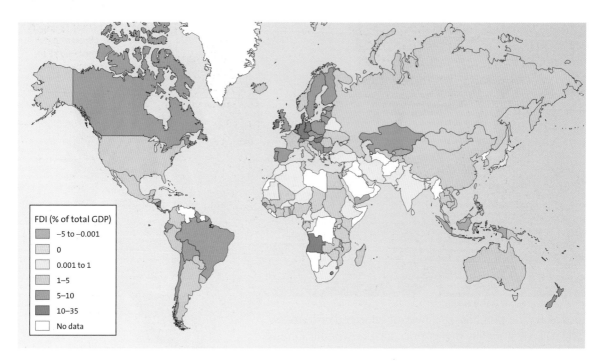

FDI (% of total GDP)
- −5 to −0.001
- 0
- 0.001 to 1
- 1–5
- 5–10
- 10–35
- No data

Figure 39 FDI as proportion of total GDP, 2000

land. The accelerating role of FDI is clearly a result of economic globalisation. There are now over 60 000 TNCs, controlling over 750 000 foreign affiliates. It can be a significant means of development for LDCs so that their commercial agriculture, resources, industries and tourism can be established or expanded. Most countries have a positive balance of growing FDI as a percentage of GDP (Figure 39). There is now a consensus among governments of industrialised and non-industrialised countries that FDI is desirable for economic growth and can reduce poverty. Critics argue that it leads to over-dependent or restrictive development, whereby TNCs control substantial 'chunks' of an economy such as oil or mining. Supporters have suggested that FDI buys capital and technology, develops skills and linkages, and increases employment and incomes.

Sources of FDI

FDI originates overwhelmingly from MEDCs. Traditionally, until the 1980s, Europe and the USA were the main sources of FDI. They were joined by Japan, and by 2000 the base had spread to include a range of Asian NICs and Brazil. Most of the world's FDI (66%) is between and into developed economies, which is actually a reversal of the colonial pattern of flows in the early twentieth century. Recently, investment has widened, with FDI being focused on a few industrialising nations in eastern Europe. Up to 70% of the FDI to developing countries went to just nine countries, with nearly 40% going to China (including Hong Kong). This leads to a highly uneven spread of recipients (see Part 4).

Historically, a major proportion of FDI was concentrated upon natural resource-based activity, such as mines and plantations, because the TNCs offered capital and technology. In the 1980s, FDI was an integral part of the global shift of manufacturing, especially in technologically more advanced sectors, and large-volume, medium-technology consumer goods such as televisions and cars. By 2000, FDI was important in the service sector, including finance, telecommunications, tourism and retailing.

- The largest flows of FDI occur between advanced economies (North America, Europe and Japan).
- Flows to developing economies are growing, with a particular concentration on China, but also with increases to all countries in Asia, Latin America and Africa.
- FDI can be seen as a way out of poverty, depending on the way the legal frameworks for FDI deals are set up.

A further important source of national income for some countries is remittances from workers living abroad. Table 22 is arranged by the scale of remittances and shows that for many LEDCs remittances are equivalent to a large volume of the country's trade in goods. Table 22 also shows the importance of proximity for migration and remittances in the cases of Mexico and the European states. In the case of Indonesia, it is proximity to Singapore and Malaysia that has attracted the workers, while Yemenis go to work in Saudi Arabia. Remittances only benefit some of the more prosperous LEDCs and not the countries that are facing the worst human disasters. Nevertheless, many people working abroad cumulatively make a major contribution to their home countries, with many returning home to set up their own small enterprises from their newly acquired wealth.

Country	US$ (billions)	% of merchandise trade
Mexico	10	6
India	8	17
Spain	4	3
Pakistan	4	36
Portugal	3	13
Egypt	3	66
Morocco	3	36
Bangladesh	3	47
Colombia	2	20
Serbia and Montenegro	2	92
Dominican Republic	2	37
Turkey	2	6
El Salvador	2	65
Jordan	2	70
Brazil	2	3
China	2	1
Guatemala	2	71
Ecuador	1	28
Yemen	1	40
Sri Lanka	1	27
Indonesia	1	2
Greece	1	11
Jamaica	1	102
Poland	1	3
Tunisia	1	16
Total world	**76**	–

Table 22 Remittances (US$) as a proportion of trade in merchandise for selected countries, 2002

Aid

This section examines the geography of aid in the light of the development gap. Official Development Assistance (ODA) is defined by the Development Assistance Committee (DAC) of the OECD as all grants and loans made by the 22 DAC countries and NGOs (non-governmental organisations) to countries that are categorised according to their eligibility for aid. The countries considered most eligible to receive aid are LDCs, LICs (low-income countries) and the LMIC/UMIC (low- and upper-middle income countries). Some former Soviet bloc countries, NICs and RICs may receive some official aid. Some NIC and RIC transitional states do give ODA, as shown by the data for South Korea, Kuwait and UAE in Table 23.

Table 23
Aid in 2002

	Aid received 2002 (US$ per capita)	Aid received 2002 (% of GNI)	Aid donated, 2002*(US$ per capita)	Aid donated, 2003 (% of GNI)	Untied aid as % of aid, 2002	Development assistance status
USA	–	–	46 (23%)	0.15	ND	Member
Japan	–	–	76 (16%)	0.20	82.8	Member
UK	–	–	78 (9%)	0.34	100	Member
Germany	–	–	60 (9%)	0.28	86.6	Member
Canada	–	–	64 (3%)	0.24	61.4	Member
France	–	–	86 (9%)	0.41	91.5	Member
Russia	9	0.4%	–	–	–	Transition
Denmark	–	–	286 (2%)	0.84	82.1	Member
Spain	–	–	38 (3%)	0.23	59.9	Member
Kuwait	2	0	–	–	–	Transition
UAE	1	0	–	–	–	Transition
Hungary	46	0.7	–	–	–	Transition
South Korea	–	0	–	–	–	Transition
Malaysia	4	0.1	–	–	–	UMIC
Brazil	2	0.1	–	–	–	UMIC
China	1	0.1	–	–	–	LMIC
Burkina Faso	40	15.2	–	–	–	LDC
Tanzania	35	13.2	–	–	–	LDC
Rwanda	44	20.8	–	–	–	LDC

*Figures in parentheses are % global aid.

Table 23 looks at the same countries as Table 2 (see p. 7) and shows two measures of the value of aid — as a percentage of GNI and as a per capita figure. Although the USA is the largest donor, contributing 23% of the global total, its contributions per capita and as a percentage of GNI are among the lowest. The contribution of European DAC members is high. Scandinavia has the highest per capita contribution of all and the highest proportions of GNI, as illustrated by Denmark.

Figure 40 shows the destination of aid from four donor countries. As can be seen, political events and past links are important in determining present-day aid. Donations from the USA in 2002 demonstrate a clear link to political events of the past decades:

■ Russia — as an emerging G8 member after 1989
■ Serbia and Montenegro — after the civil wars in the former Yugoslavia
■ Afghanistan — after 9/11
■ Israel and Egypt — because of the crisis in the Middle East over the past 50 years and because of the strength of the Jewish lobby in the US political system

In contrast, the UK and France illustrate the neocolonial aspect of aid donation:

- India, Ghana and Tanzania are all former British colonies and are current members of the Commonwealth. Sierra Leone and South Africa are other major recipients of aid from the UK.
- Cote d'Ivoire and Cameroon were French colonies, as were other major recipients such as Senegal and Vietnam.

Japanese aid has a more striking regional dimension, being primarily focused on east and south Asia (70%).

Flows of aid might be distorted when countries are affected by disasters such as the 2005 tsunami, the Pakistan earthquake and hurricane Katrina — countries from all around the world, including many LEDCs, offered aid to Louisiana. It is not surprising, therefore, that European countries that have been campaigning for aid and debt relief in sub-Saharan Africa are trying to keep the aid needs of Africa in the public's mind. The 'Make Poverty History' movement, which emerged from the UN Johannesburg Conference (2002), was the focus of the G8, a UN General Assembly summit looking at progress in meeting the 2000 Millennium Development Goals (see Part 2) and the WTO Congress in Hong Kong, all in 2005. 'Charity fatigue' is a well-analysed phenomenon.

The Monterrey UN Conference in 2002 agreed to scale-up aid to developing economies to help them achieve the Millennium Development Goals. The Millennium Goal of ODA reaching 0.44% of GNI in 2006 is not being met.

Figure 40
The destinations of aid from the USA, France, the UK and Japan, 2002

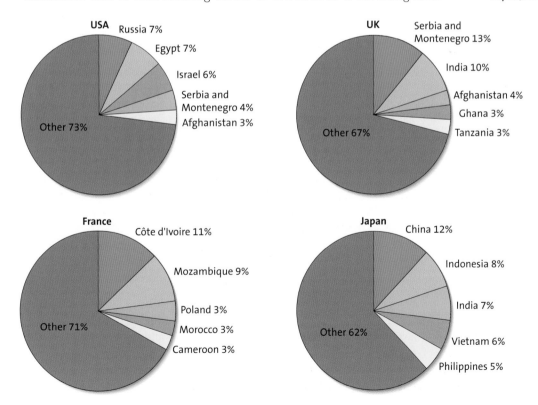

USA
Russia 7%
Egypt 7%
Israel 6%
Serbia and Montenegro 4%
Afghanistan 3%
Other 73%

UK
Serbia and Montenegro 13%
India 10%
Afghanistan 4%
Ghana 3%
Tanzania 3%
Other 67%

France
Côte d'Ivoire 11%
Mozambique 9%
Poland 3%
Morocco 3%
Cameroon 3%
Other 71%

Japan
China 12%
Indonesia 8%
India 7%
Vietnam 6%
Philippines 5%
Other 62%

The role of non-governmental organisations (NGOs)

Non-governmental organisations (NGOs) collectively provide more aid than the World Bank. There has been an explosion in the number of NGOs, from 1600 in 1980 to nearly 10 000 by 2000. The term NGO is a bit of a misnomer because most organisations depend, in some measure, on government funds. Concern has been expressed that the very large US NGOs or big international NGOs (BINGOs) have become too powerful and overly bureaucratic. For example, the American Red Cross has an income greater than Mali.

There is no such thing as a typical NGO — they vary in size, operating style, geographic focus, religious background and programme orientation. Some have developed from a religious foundation responding to a need, or from a single individual's motivation. Examples of organisations with multi-million dollar budgets and large numbers of staff include World Vision and Oxfam. Programme orientation varies from those with a single focus, such as Water Aid, Médecins Sans Frontières and Farm Africa, to multi-purpose organisations, such as CARE.

NGOs have three functional roles:
- disaster relief
- technical assistance (now called Practical Action)
- network and institution building — for example, the 'Trickle Up' programme, which facilitates indigenous enterprises

Over time, NGOs evolve from emergency disaster relief organisations, through to sustainable development managers. Ultimately, they facilitate the takeover of their work by LDC citizens and so can be temporary organisations in a particular location.

The strengths of NGOs are their 'bottom-up' approach and their ability to provide tailor-made, locally appropriate projects. Many comparatively small institutions tend to be less bureaucratic and potentially more cost-effective. As apolitical organisations, they have the ability to reach a broader audience and are less contentious than bilateral government aid and investment (FDI). Well-organised NGOs often employ specially trained and locally based staff to ensure appropriateness, and overall have a good image with donors because they are perceived as cost-effective and less corrupt.

NGOs collaborate with multilateral institutions and some, such as WWF, play a considerable role in moulding policy. Recently, NGOs have become 'the darling of the aid industry', administering a substantial percentage of government aid. NGOs play an increasing role in development education — for example, Amnesty International's work on human rights. Oxfam's campaign on restructuring debt in the late 1990s is another excellent example and has had a major impact at G8 summits.

Since 1970, NGOs based and developed 'south' of the Brandt line, often known as grassroots groups or base groups, have proliferated. Examples are:
- the Grameen Bank of Bangladesh — involved in microenterprise development
- the Self-employed Women's Association (SEWA), Ahmedabad, India — which grew out of a trade union for women street traders
- Ghanaian cocoa farmers cooperative — operating in farming areas

For effective operation, it is vital that northern and southern NGOs cooperate and that they work in partnership with governments. There can be political conflicts because NGOs can empower and may incite local people in politically sensitive parts of the country (e.g. Ecuadorean ecotourism). Their efficiency can even be a threat to governments.

Sometimes, locally based projects can be incorporated into a national framework. *Case study 12* explores the work of CAFOD in African countries.

CAFOD IN AFRICA

Case study **12**

In 2003–04, CAFOD (the Catholic Agency for Overseas Development) spent over £12 million in Africa. CAFOD is aiding a selection of African countries (Figure 41). There are many reasons for their work on the continent:

- CAFOD focuses on the 18 lowest-ranking countries in the UN **HDI** in Africa, which assesses life expectancy, education and standard of living.
- The benefits of globalisation (for example, improved communications and trading opportunities) are only felt by a limited few in Africa. Foreign investors often have more economic influence than local communities or even governments.
- Long-lasting wars and civil conflicts (e.g. in Sierra Leone), have blighted the development of many countries, deepening poverty and uprooting civilians. Competition to exploit natural resources has fuelled many conflicts. Children are forced to fight in some wars (e.g. in Liberia) and the targeting and violation of women (e.g. in Rwanda and Burundi) have become military tactics, leaving a deep legacy of trauma.
- The debt crisis and unfair international trade rules have further deepened poverty and widened the gulf between rich and poor.
- Over 28 million people are living with AIDS and HIV, placing a huge strain on health services. The UN estimates that AIDS could kill up to 26% of the labour force in the worst-affected countries by 2020. By 2000, there were an estimated 12.1 million children orphaned by AIDS.
- A lack of economic and social justice — including respect for human and civil rights, good governance, freedom of speech and the opportunity to take part in democratic processes — is the reality in some countries (e.g. Zimbabwe).
- Many countries are prone to drought (e.g. Niger, 2006) and floods and the spread of El Niño events. Schemes such as dams and large-scale mechanised agriculture have caused environmental degradation.
- Much of Africa's poverty is linked to a history of exploitation, including slavery, colonialism and **neocolonialism**, which have left a bitter legacy.

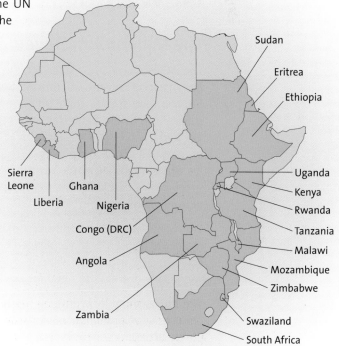

Figure 41 *Countries in Africa receiving aid from CAFOD*

One country that benefits from CAFOD aid is the Democratic Republic of the Congo. CAFOD provided DRC with £550 000 in 2004. The money was spent on dealing with the aftermath of the arrival in the Goma area in 1994 of 1 million refugees from Rwanda and from the subsequent civil war, which displaced 2 million people. Food is scarce and 33% of the population are short of basic nutrition. Malnutrition affects half of children under the age of five.

The work of CAFOD is partly supported by the UK government. Much of the emphasis is on education of the refugees and schooling, so that the people can re-establish their lives. Nevertheless, the aid from this one NGO will only scratch the surface of the problems around Goma.

Websites: **www.cafod.org.uk**
www.itgd.org
www.msf.org/msfinternational/aboutmsf
www.ngos.net

17
Using case studies

Question
What are the benefits to organisations such as CAFOD of focusing aid?

Guidance
Consider whether small-scale projects:
- are helpful
- have sufficient funds
- have adequate infrastructure

New approaches

The Millennium Development Goals

The Millennium Declaration of the UN drew attention to countries where the proportion of income or consumption of the poorest 20% is very low. Even in the UK, the poorest 20% share only 6.1% of national wealth; in Japan, the same group shares 10.6%; the figure for the Czech Republic is 10.3%. In contrast, the figures for Sierra Leone (1.1%), Lesotho (1.5%), Namibia (1.4%) and South Africa (2%) suggest that economic poverty affects significant numbers of people. The data are for whole countries, some of which have very affluent minorities, often located in specific regions. Economic wellbeing within countries can vary as much as it does between countries.

The Millennium Development Goals (MDGs) are outlined in Table 24. All 192 members of the UN have pledged to meet these goals by 2015. Recognising that economic growth alone will not allow many LDCs to bridge the development gaps is crucial — hence the need for the MDGs, which are global targets.

The MDGs show how poverty has a major impact on the many aspects of development. Development is now measured across a wider range of criteria, rather than purely economic ones, including aspects of the quality of life, such as health and

Goal	Target
1 Eradicate extreme poverty and hunger	• Halve, between 1990 and 2015, the proportion of people whose income is less than $1 a day • Halve, between 1990 and 2015, the proportion of people who suffer from hunger
2 Achieve universal primary education	• Ensure that, by 2015, children everywhere, boys and girls alike, will be able to complete a full course of primary schooling
3 Promote gender equality and empower women	• Eliminate gender disparity in primary and secondary education, preferably by 2005, and in all levels of education no later than 2015
4 Reduce child mortality	• Reduce by two-thirds, between 1990 and 2015, the under-five mortality rate
5 Improve maternal health	• Reduce by three-quarters, between 1990 and 2015, the maternal mortality ratio
6 Combat HIV/AIDS, malaria and other diseases	• Have halted by 2015 and begun to reverse the spread of HIV/AIDS • Have halted by 2015 and begun to reverse the incidence of malaria and other major diseases
7 Ensure environmental sustainability	• Integrate the principles of sustainable development into country policies and programmes, and reverse the loss of environmental resources • Halve, by 2015, the proportion of people without sustainable access to safe drinking water and basic sanitation • Have achieved, by 2020, a significant improvement in the lives of at least 100 million slum dwellers
8 Develop a global partnership for development	• Develop further an open, rule-based, predictable, non-discriminatory trading and financial system (includes a commitment to good governance, development, and poverty reduction — both nationally and internationally) • Address the special needs of the least developed countries (includes tariff- and quota-free access for exports, enhanced programme of debt relief for HIPCs and cancellation of official bilateral debt, and more generous ODA for countries committed to poverty reduction) • Address the special needs of landlocked countries and small island developing states • Deal comprehensively with the debt problems of developing countries through national and international measures, to make debt sustainable in the long term • In cooperation with developing countries, develop and implement strategies for decent and productive work for youth • In cooperation with pharmaceutical companies, provide access to affordable, essential drugs in developing countries • In cooperation with the private sector, make available the benefits of new technologies, especially information and communications

Table 24 *The UN Millennium Development Goals*

Millennium Development Goal		Low-income fragile states	Other low- and middle-income states
	Population	871 million	4361 million
MDG 1	Number of people living on <$1 per day Proportion undernourished (mean 1999–2001)	343 million 33%	821 million 15%
MDG 2	Primary education enrolment	70%	86%
MDG 3	Primary education enrolment ratio, boys:girls	0.84	0.92
MDG 4	Child mortality rate per 1000 (2002)	138	56
MDG 5	Maternal mortality rate per 100 000	734	270
MDG 6	Number of people living with HIV/AIDS (2001) Malaria death rate per 100 000	17.1 million 90	21.4 million 7
MDG 7	Proportion of population without access to safe water	38%	18%
MDG 8	Telephone and cellphone subscriptions per 100 people	4.5	18.8

Table 25 *Progress on Millennium Development Goals in fragile states and other poor countries, 1990–2000*

education. Recent assessments of progress towards the MDGs confirm the enormous problems faced by the world's poorest countries for whom the development gap has truly widened and whose future might be affected by climate change.

The concept of 'fragile states' illustrates the challenge of meeting the MDGs (Table 25). The 46 states listed by the Department for International Development (DFID) as fragile states are failing relative to other low-income states against all the goals listed in Table 24.

18 Using case studies

Question

So, just 6 years on, what progress is being made?
(a) Try to explain why some goals are more difficult to achieve than others. Guidelines might include the scale and difficulty of the goal (e.g. Goal 1) or extraneous factors, such as the impact of HIV/AIDS on child mortality or the scale of finance required to fulfil it.
(b) Study Table 25. Suggest reasons why the so-called fragile states are failing to achieve the MDGs even compared with other low -income nations.

The UN Global Compact 2000

Increasingly, various sectors and organisations are working together to tackle the world's greatest problems. The Global Compact between the UN and its six agencies attempts to persuade companies to take a sustainable approach to their activities. The Compact involves governments, companies, labour, society organisations and the UN, with the UN as a convenor and facilitator. The Compact has ten principles for companies to act on in their own corporate environment. The principles are shown in Table 26.

Companies are not expected to address all ten principles at once. Considering the vast differences in size and location of companies around the world, employers are encouraged to work in their own way and at their own speed towards enacting these principles in the workplace. There are 1842 member companies.

Table 26 *Principles of the Global Compact*

Human rights
1 Support and respect the protection of international human rights within their sphere of influence
2 Make sure their own corporations are not complicit in human rights abuses
Labour
3 Freedom of association and the effective recognition of the right to collective bargaining
4 Elimination of all forms of forced and compulsory labour
5 The effective abolition of child labour
6 Elimination of discrimination in respect of employment and occupation
Environment
7 Support a precautionary approach to environmental challenges
8 Undertake initiatives to promote greater environmental responsibility
9 Encourage the development and diffusion of environmentally friendly technologies
Anti-corruption
10 Businesses should work against corruption, including extortion and bribery

Bayer, Germany, signed the Global Compact at its founding meeting in 2000. The company's annual report has a special section on the Global Compact. To document its commitment to the principles of the Compact, Bayer has four examples:

- financial support for a Brazilian foundation for the rights of the child, which combats child labour
- donations of profits from two pharmaceutical products against sleeping sickness to the World Health Organization
- efforts to control the spread of antibiotic resistance
- company training programmes for Brazilian farm workers and small farmers, teaching them to handle pesticides appropriately

The fashion retailer Hennes and Mauritz (H&M) has addressed the issue of child labour in Bangladesh. There is a demand for child labour in Bangladesh because clothing suppliers know that these low-skilled people are cheap to employ and easy to dismiss. In 1999, H&M set up a social responsibility policy to provide vocational training for former child labourers. The 56 children on the programme have their wages paid by H&M for 7 months while they have theoretical and practical training before they go to guaranteed jobs as machine operators in one of the suppliers. The scheme has improved skill levels while also making the students more socially aware. It has also increased awareness of the value of education among the families of the trained workers.

Working for Africa: international efforts in the new millennium

2005 was notable both for the G8 Gleneagles Conference and the Live 8 concerts that publicised Africa's development problems. However, concern for Africa had been voiced before.

- The New Partnership for Africa (NEPAD) was founded in 2001 by the African Union to campaign for more funds for science, technology and education in Africa.
- The 17-person Commission for Africa, founded in 2004, tries to raise funds for university centres of excellence in African countries. Only 83 persons per million work in science, technology and R&D in Africa, compared with 1102 per million in MEDCs.

The G8 agreed to assist the African continent in the following ways:

- By doubling aid by 2010 — an extra $50 billion worldwide and $25 billion for Africa.
- Writing-off immediately the debts of 18 of the world's poorest countries, most of which are in Africa (Table 27). This is worth $40 billion now, and as much as $55 billion as more countries qualify.
- Writing off $17 billion of Nigeria's debt, in the biggest single debt deal ever.
- A commitment to end all export subsidies, probably by 2010, should be agreed by the WTO. The G8 has also committed to reducing domestic subsidies, which distort trade.
- Developing countries will 'decide, plan and sequence their economic policies to fit with their own development strategies, for which they should be accountable to their people'.
- As close to universal access to HIV/AIDS treatments as possible by 2010.

- Funding for treatment and bed nets to fight malaria, saving the lives of over 600 000 children every year.
- Full funding to totally eradicate polio from the world.
- By 2015, all children to have access to good-quality, free and compulsory education and to basic healthcare, free where a country chooses to provide it.
- Up to an extra 25 000 trained peace-keeping troops, helping the Africa Union to respond better to security challenges, such as Darfur.

Table 27 *HIPCs and eligibility for debt relief (African countries in red)*

Eligible in 2005		Probably added shortly	Not yet judged eligible by G8
Benin	Mauritania	Cameroon	Burundi
Bolivia	Mozambique	Chad	Central African Republic
Burkina Faso	**Nicaragua**	Democratic Republic of the Congo	Comoros
Ethiopia	Niger	The Gambia	Republic of the Congo
Ghana	Rwanda	Guinea	Côte d'Ivoire
Guyana	Senegal	Guinea-Bissau	**Laos**
Honduras	Tanzania	Malawi	Liberia
Madagascar	Uganda	São Tomé and Principe	**Myanmar**
Mali	Zambia	Sierra Leone	Somalia
			Sudan
			Togo

The following commitments were made by individual countries as a result of Gleneagles:
- *EU* — pledge to reach 0.56% ODA as a proportion of GNI by 2010 and 0.7% ODA/GNI by 2015; 50% of increase in ODA will go to sub-Saharan Africa
- *Germany and Italy* — 0.51% by 2010 and 0.7% by 2015
- *France* — 0.5% by 2007, 0.7% by 2012, so aid to be doubled by 2007 and 66% for Africa
- *UK* — 0.7% by 2013; aid to double by 2008
- *USA* — double aid to Africa by 2010 but no % of ODA/GNI stated
- *Japan* — measures to double ODA to Africa and improve health provision
- *Canada* — has an investment fund for Africa and will double aid by 2010 as well as funding disease prevention; other funds for Darfur and to support humanitarian work
- *Russia* — funds to cancel debt and provide for debt relief to HIPCs

In 2006, the Department for International Development in the UK set out plans to provide long-term finance for education in the poorest countries in Africa so that the goal of free schooling for all is a reality by 2015.

Case study 13

THE MAKE POVERTY HISTORY CAMPAIGN MESSAGE

Make Poverty History urges governments and international decision makers to rise to the challenge of 2005. We are calling for urgent and meaningful policy change on three critical and inextricably linked areas: trade, debt and aid.

Trade justice
- Fight for rules that ensure governments, particularly in poor countries, can choose the best solutions to end poverty and protect the environment. These will not always be free-trade policies.

Contemporary Case Studies

- End export subsidies that damage the livelihoods of poor rural communities around the world.
- Make laws that stop big businesses profiting at the expense of people and the environment.

The rules of international trade favour the most powerful countries and their businesses. On the one hand, these rules allow rich countries to pay their farmers and companies subsidies to export food — destroying the livelihoods of poor farmers. On the other hand, poverty eradication, human rights and environmental protection come a poor second to the goal of 'eliminating trade barriers'.

We need trade justice, not free trade. This means the EU single-handedly putting an end to its damaging agricultural export subsidies now; it means ensuring poor countries can feed their people by protecting their own farmers and staple crops; it means ensuring governments can effectively regulate water companies by keeping water out of world trade rules; and it means ensuring trade rules do not undermine core labour standards.

We need to stop the World Bank and the IMF forcing poor countries to open their markets to trade with rich countries, which has proved so disastrous over the past 20 years; the EU must drop its demand that former European colonies open their markets and give more rights to big companies; we need to regulate companies — making them accountable for their social and environmental impact both here and abroad; and we must ensure that countries are able to regulate foreign investment in a way that best suits their own needs.

Drop the debt

- The unpayable debts of the world's poorest countries should be cancelled in full, by fair and transparent means.

Despite grand statements from world leaders, the debt crisis is far from over. Rich countries have not delivered on the promise they made in the twentieth century to cancel the debts of poor countries. Many countries still have to spend more on debt repayments than on meeting the needs of their people. Rich countries and the institutions they control must act now to cancel all the unpayable debts of the poorest countries. They should not do this by depriving poor countries of new aid, but by digging into their pockets and providing new money.

The task of calculating how much debt should be cancelled must no longer be left to creditors concerned mainly with minimising their own costs. Instead, we need a fair and transparent international process to make sure that human needs take priority over debt repayments.

International institutions like the IMF and World Bank must stop asking poor countries to jump through hoops in order to qualify for debt relief. Poor countries should no longer have to privatise basic services or liberalise economies as a condition for getting the debt relief they so desperately need.

To avoid another debt crisis, poor countries need to be given more grants, rather than seeing their debt burden piled even higher with yet more loans.

More and better aid

- Donors must now deliver at least $50 billion more in aid and set a binding timetable for spending 0.7% of national income on aid. Aid must also be made to work more effectively for poor people.

Poverty will not be eradicated without an immediate and major increase in international aid. Rich countries have promised to provide the extra money needed to meet internationally agreed poverty reduction targets. This amounts to at least $50 billion per year, according to official estimates, and must be delivered now. Rich countries have also promised to provide 0.7% of their national income in aid and they must now make good on their commitment by setting a binding timetable to reach this target.

However, without far-reaching changes in how aid is delivered, it won't achieve maximum benefits. Two key areas of reform are needed. First, aid needs to focus better on poor people's needs. This means more aid being spent on areas such as basic healthcare and education. Aid should no longer be tied to goods and services from the donor, so ensuring that more money is spent in the poorest countries. The World Bank and the IMF must become fully democratic in order for poor people's concerns to be heard.

Second, aid should support poor countries' and communities' own plans and paths out of poverty. Aid should therefore no longer be conditional on recipients promising economic change like privatising or deregulating their services, cutting health and education spending, or opening up their markets: these are unfair practices that have never been proven to reduce poverty. Aid needs to be made predictable, so that poor countries can plan effectively and take control of their own budgets in the fight against poverty.

Websites: www.food4africa.org.za
www.dfid.gov.uk

19 **Using case studies**

Question

Assess some of the ways in which Third World debt could be reduced.

Guidance

- Read the summary of the Make Poverty History campaign message and assess how useful each solution is in making poverty history. This can then be noted and used as an essay plan.
- This question asks you to 'assess'. Therefore, you should mention at least three ways, because you will need to say which way is the most effective and justify why. It is best to assess at the end, because it reminds the examiner that you have followed the directions in the question.
- 'Debt relief' is the current favourite, but you need to note both the positives and the drawbacks for countries that are unable to comply with the conditions.
- A discussion of trade and fair trade would be useful.

Economic futures

In this part we look to the future, at how work and employment might change, and then at how economic activities need to become more sustainable (the concept of 'green growth').

The future of work

Work has changed over the centuries and will continue to do so.
- Work only became a paid occupation in the eighteenth century.
- Work became more specialised during the Industrial Revolution, with a division of labour between tasks.
- The technologies of the nineteenth century tied people to factory working and the exploitative control of the labour force by the owners.
- Assembly lines in the early twentieth century revolutionised production, but also led to increased collective action supported by trades unions.
- After 1918, much production was run and supported by increasing numbers of administrators, as producer services grew in importance.
- By 1956, white collar work had overtaken blue collar work in the USA, and Europe was to follow soon after.
- In the late twentieth century, manufacturing employment declined as service occupations grew, aided by IT developments.

Other factors are influencing attitudes to work and the workplace. For instance, inheritance is providing some people with the opportunity to curtail or switch their career or to move abroad. Harassed employees are beginning to rank 'quality of lifestyle' as the key focus of their lives.

It is difficult to predict the future. In the 1980s, people writing about the future development of the world mentioned Japan, but not India or China. People were looking towards more leisure hours and fewer working hours. This trend has continued in some countries, such as France, but in the USA, Japan and the UK, many have increased their working hours.

There have been three major trends that have impacted on the world economy:
- the entry of China and India into the world economy — bringing 2.5 billion producers and consumers (38% of the world's population)

- technology — enabling companies to distribute their work around the globe and be open for business 24/7
- the ageing and declining numbers of workforce in many MEDCs

Today, the shift is towards a knowledge-based economy rather than a product- and service-based economy. Many MEDC cities, such as London, already have more people employed in knowledge industries, such as R&D, education at all levels and consultancy, than the conventional 'city' jobs in finance and business. 70% of the costs for a developed world company are labour costs and only 30% are capital. It is inevitable that rising labour costs and declining capital costs lead to outsourcing.

Changes in the workplace

Computers in the workplace were almost unknown 50 years ago. At first they were vast main-frame machines and only since the 1970s have they become the desk tops that dominate offices today. Computing continues to alter the way we work. Laptops and WiFi hot spot technology are already enabling 'work' to take place on the move, while working at home (teleworking), often linked to a move to a rural environment, is a common choice. The technology to enable information to be shared is becoming increasingly sophisticated. The outcome is predicted to be more outsourcing, with less need for city centre office clusters. Work can be carried out anywhere and at any time. The rigid working hours of shift work and/or 9 to 5 working are being eroded. The Intel working pattern (see *Case study 1*, p. 20) expects managers to engage in teleconferencing outside of the working day because of the time zone spread of the company's workforce.

Changes in the size of the workforce in MEDCs

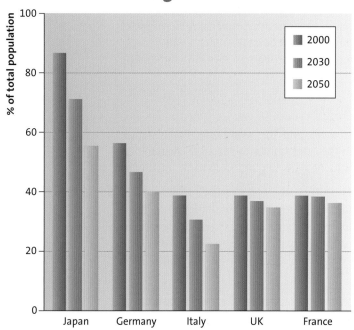

Figure 42 *Projections of working age population in Japan and western Europe*

The impact of low birth rates in MEDCs is beginning to affect the job market. More people are retiring from work than are joining the labour force. Even the phenomena of the gap year (a sabbatical year to travel the world) and gardening leave (taking time off between jobs) are having a relatively small effect on the size of the labour force.

Figures 42 and 43 show the nature of the workforce crisis:
- in western Europe and Japan a declining workforce is a reality (Figure 42)
- in the USA it will be a reality by 2030 (Figure 43)
- Australia, New Zealand, Canada and much of the rest of the world continue to have larger work forces fuelled by immigration

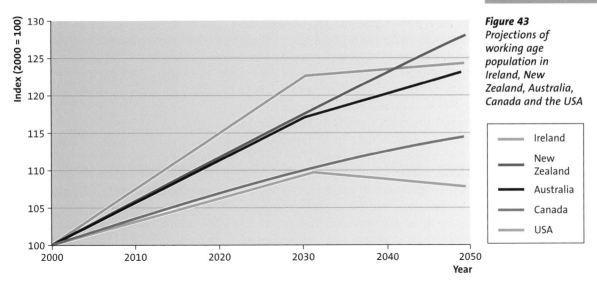

Figure 43
Projections of working age population in Ireland, New Zealand, Australia, Canada and the USA

Legend:
- Ireland
- New Zealand
- Australia
- Canada
- USA

If migration is permitted, it is possible for economic migrants, refugees and asylum seekers to help plug the shortage of workers and skills, as has been suggested in the UK in 2006. However, many right-wing political parties oppose migration as a solution, partly appealing to the votes of the less-qualified, less-adaptable voters whose jobs are most threatened by migrant workers.

Importing workers, rather than outsourcing jobs (see Part 6), alters the demographic profile of a country. There are more people filling the demographic gap in the workforce and also paying taxes, which may help to support the elderly. The young age of the immigrants may lead to higher birth rates. In the 1990s, the USA attracted 200 000 skilled workers, who filled gaps in the labour force in IT. However, since 9/11, the restrictions on skilled immigrants are having the opposite effect and are causing more companies to send IT work offshore. Nevertheless, there is an enormous migration of both legal and illegal Mexican migrants to work in low-skilled jobs in the USA. Many of these actually settle in California and Texas.

Zacatecas is a state of Mexico that was once dependent on silver mining. Today, 800 000 people from the state (40% of its population) live in the USA. The impact is that the state is rich because of remittances. The state has encouraged 'three for one' schemes, so that every dollar remitted by emigrants to the USA is matched by state and municipal dollars for approved schemes in the puebla of Juriquilla. The result is over 1000 infrastructural and water projects. The consequence is improved quality of life for those left behind, who are increasingly the elderly in the remote pueblos. Remittances do not stimulate the local economy. Therefore, filling the labour gap in a developed country is benefiting both the USA and Zacatecas. However, in 2006, a move to 'legalise' all illegal migrants to the US and accept the status quo was facing strong opposition, especially from lower-income Americans.

Women in the labour force

The increasing proportion of women in the labour force has been a feature of MEDCs over the past 30 years. However, the recruitment of women will not continue to fill the demographic gap because it is difficult to raise the levels of female employment further without affecting the birth rate. The traditional role of women as child raisers has taken many out of the labour market. In the twenty-first century, governments are more aware of the needs of women who want a 'work–life balance'. The traditional solution has been to encourage part-time work, but this is generally less well paid. Part-time work does not help parenting units if, until the arrival of a child, there were two incomes to support the household (the DINKIs — dual income, no kids). Job-sharing is another solution. More recently, support for child minding, either directly or through the tax system, has been introduced to enable mothers to return to work. This may help some but, when child care can cost £300 a week in London in 2005, it might only be the wealthy who gain from the subsidies. The demographic solution to this issue is the increasing use made by many of grandparents. This group may have an increasing role to play when children are at school and need collecting during the afternoon. After-school clubs are a further solution that enables both parents to remain at work.

Table 2 on p. 7 shows that women are a relatively constant proportion of the labour force in most countries, although the table masks the different nature of female labour and the reasons for that. The lowest proportions are found where there are cultural constraints, such as those found in Islamic societies (e.g. in the United Arab Emirates). The highest proportions are found where the egalitarian principles of socialist regimes have encouraged women to work — for example, Russia and Tanzania. The high proportion of women in the labour force in African states is due to the role of women in agriculture and, possibly, to the impact of HIV/AIDS on the male population.

Table 2 also shows that female work is predominantly in the service sector. With the exception of the HIPCs, male employment is also dominated by service employment. The proportion of male service sector employment indicates the approximate level of development.

Changes in RICs as they become NICs

This process is best illustrated by China, which was designated an NIC in 2005.

Case study 14   **THE PEOPLE'S REPUBLIC OF CHINA: THE WAKING GIANT**

The current growth of the Chinese economy began in 1978 when the country launched economic reforms in the southern economic zone of Guangdong, adjoining Hong Kong. Hong Kong was then a separate NIC. Special economic zones were the main strategy for attracting inward investment and there are now over 6000. In 2001, China entered the WTO — confirming its status as a significant force for change in the global economy.

In what ways is China changing the world pattern of economic development? In some markets, such as steel, coking coal, the shipping of raw materials to China and soya beans, Chinese decisions about the price that it will pay tend to set the global trading price. China's trade in raw materials from Brazil, Argentina and Australia is growing and China is now the biggest trading partner of the EU, replacing the USA in 2004. The USA fears the migration of American jobs to the Pearl River delta (Guangzhou and Shenzhen) and the Yangtze River delta (Shanghai) (Figure 44). In 2004, Spaniards

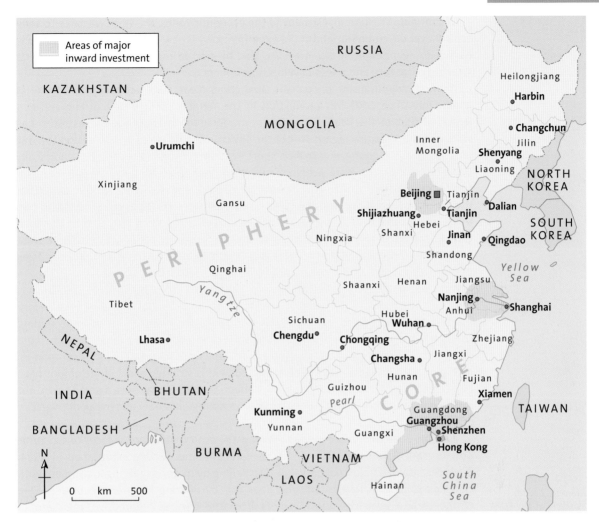

Figure 44 *China, showing the areas with major inward investment*

incinerated cheap Chinese shoes in protest against the fact that the Chinese product was undercutting Spanish-produced shoe prices. Production costs in China are 10% of those in Europe. GDP growth is approximately 9%, industrial output is growing by 10% and more per year, and agricultural output is growing by 2.5%. GDP per capita is still low by world standards ($1220) but increasing by 10% per year (all data from 2004). The country had a trading surplus of $44 billion in 2003. Many people see the huge Chinese demands for oil as a prime reason for the high oil prices.

The Chinese boom is occurring mainly because the state decided to take the direction of industrialising for export rather than following the path of the former socialist system of industry satisfying internal needs. That is not to say that there is not a huge internal market for goods. From the population of 1.3 billion people, 300 million are expected to move to the industrial centres by 2020. This shift dwarfs the movement from the countryside, where most Chinese still live, to towns that occurred in the European Industrial Revolution. Some argue that the quality of life of workers will be a sweatshop style, similar to conditions in Victorian England. The state controls exchange rates, which have kept raw material prices low, and low interest rates have enabled investment to be made cheaply.

By confining Western investment to the economic zones, the state was able to keep control over development and could experiment with new policies in a geographically confined environment. It did, however, lead to marked regional disparities — with a strong coastal core and a less-developed periphery.

The first industrialising region was Guangdong/Pearl River delta, adjacent to Hong Kong. Only since 2003 has production in the Yangtze delta/Shanghai/Pudong area exceeded that of the Pearl delta. The prime focus of foreign direct investment has shifted to the Shanghai region. Part of the shift is explained by overheating and the lack of labour in Guangdong — 2 million short in 2004. Growth had become economically unsustainable unless more people migrated to the region. The province is now trying to alter its industrial structure by moving out of light manufacturing to other areas that use more skilled labour and has tried to attract more skilled employment and less unskilled manufacturing. Toyota, Honda and Nissan have been attracted to the area. Guangdong exports 50% of Chinese electronics products by value. Another way to attract more technically sophisticated activities has been to stipulate that only firms paying 30% more than the minimum wage will be permitted to open.

A further strategy is to attract inward investment from small firms. Xiaolan has pursued this strategy since 1996. It invests jointly with foreign investors, such as lock maker Yale. The city and the industrial companies are members of a single trading company that is able to obtain government tax incentives and rebates much more quickly. The company also handles the exports and imports for all the new companies. In 8 years, 20 companies have been attracted by this strategy and a range of jobs have been created as a result of the multiplier effect. Shenzhen was once a small town, but is now one of China's richest cities. The Pearl delta region is now an emergent megacity of 40 million people.

With the 2008 Olympics coming to Beijing, the capital city is now a focus of economic development. The Tianjin Economic Development Area (TEDA), established in 1984, is a self-contained industrial park linked to Beijing by motorway. It has to compete with other parks to gain investment and does this by offering long-term subsidies. One of the earliest investors was Motorola from the USA. SEW-Eurodrive, the German maker of industrial drive motors, was another. The investors were attracted because of corporate tax concessions, tax-free profits for the first 2 years, labour-cost concessions and the connections to Beijing. When SEW opened a second plant near Shanghai and another in Beijing, the investment incentives offered by TEDA were improved.

The jewel in most parks' crown is to attract biotechnology companies. TEDA attracted a Danish biotechnology company and hopes that this company, which already has 40% of the world market for industrial enzymes, will enable it to attract more high-technology companies.

Xiamen, China: Workers at the EUPA Appliance Factory

Topfoto

Contemporary Case Studies

Question

List the factors underpinning the current growth of the Chinese economy.

Guidance

This question is asking you to extract factors from the account above. Can you find five?

As China industrialises, further moves are planned to encourage the spread of wealth inland from the coast (see Figure 44). This case study emphasises the enormous impact that the development of China as an industrial country will have in the future (see Part 4). Add India, and many other populous countries such as Pakistan, Indonesia and Bangladesh, into the equation and you can speculate further about the future impact of globalisation and the global shift.

Table 28 evaluates the winners and losers if the economic expansion of China continues at its present pace.

Table 28 Winners and losers from the economic expansion of China

	Winners/benefits	Losers/costs
China	• Chinese entrepreneurs, benefiting from the move from centralised command economy to a controlled market economy • The state's ability to invest in infra-structure and further growth • Creating an urban society • Numbers living in poverty reduced from 490 million in 1981 to 18 million in 2004 • Business tourism • Olympic Games 2008 • Beijing as a rising political and financial centre	• Regions suffering from overgrowth and unable to house migrants • Peripheral regions with little growth • Environmental costs and pollution, e.g. brown cloud • Energy demands, e.g. leading to Three Gorges Dam and flooding of 110 000 ha, displacing 1.2 million people • Women out at work seen by some as affecting children • Few international financial services because banks find regulations too restrictive, and corruption in Shanghai
Japan and Far East NICs	• Regional investment opportunities • Destination for exports • Less pressure on own environment • Hong Kong still retains rights and liberties that enable it to benefit from being part of China • It is the entry port for the Pearl River	• Job expansion threatened • Raw material costs rise due to China's demands being met • Smog blown from Guangdong into Hong Kong • Tokyo could lose out as a financial centre to Hong Kong and Shanghai
MEDCs	• Greater profits from investments due to low labour costs • Cheaper consumer goods • Industry to shift more to R&D and high value (e.g. luxury boats in the UK), fashion/designer items (e.g. Italian fashion goods)	• Job expansion slowing or declining as jobs exported to China • Raw material costs, e.g. steel, rise due to China's demands being met — will result in higher costs for MEDC producers
LEDCs	• Supply of raw materials such as oil from Sudan and minerals from Zimbabwe • Political support for LEDCs against G8	• Even less able to attract investment because of skills deficit, civil wars, disease • Share of world trade will drop further from current 0.5%. • Copying intensive rice growing in west Africa, leading to increased water demands • Currencies losing value • Increased incentive to migrate

The impact of the flexible workforce

In the future, what will work be like in MEDCs? Indeed, will there be any work left for MEDCs as competition grows in all sectors?

The future of work will see a flexible labour market, with reduced unionisation, and a move away from quality benefits such as healthcare and company pensions. Equally, many workers will not have a career for life but a mosaic of short-term contracts. Interestingly, unionisation was identified as a key factor in the last century with its power to give security and safe working conditions to vulnerable workers. The demands of a flexible workforce have added to stress for many workers. Equally, as governments move to cut costs and companies seek to be more efficient, this has led to the dismantling of the universal health and welfare benefits, with a move to targeting and self-provision. Many LEDCs have high levels of workforce insecurity, with many workers in Asian, African and even eastern European countries expecting poverty in old age in the absence of well-developed pension and welfare systems.

One of the main changes caused by globalisation is greater economic insecurity for many of the world's people. Figure 45 shows how the International Labour Organization has ranked countries by the security their workers enjoy in terms of income, union representation, safety at work, healthcare and social security, using an ESI (economic security index).

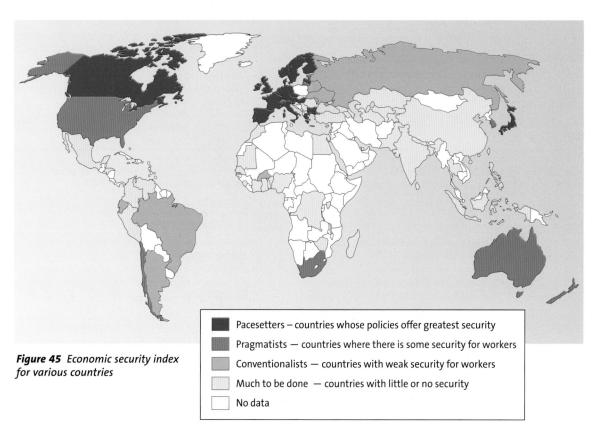

Figure 45 *Economic security index for various countries*

Pacesetters – countries whose policies offer greatest security

Pragmatists — countries where there is some security for workers

Conventionalists — countries with weak security for workers

Much to be done — countries with little or no security

No data

Question

Study Figure 45, which shows variations in the ESI. Describe and suggest reasons for the variations shown.

Guidance

Note that you have to describe *and* give reasons.

Green growth: the only future?

'Green growth' is the term used to denote more sustainable economic activities. All productive wealth-creating activities are polluters through the production process and, indirectly, through their products (e.g. plastic bags). The challenge for decision makers in organisations and companies is to remain commercial, yet manage production so that it does not degrade the environment, deplete resources in an unsustainable manner or exacerbate the growing inequality between nations and people — in particular, providing equitable lives for the world's poorest people. Economic activities have the power to both degrade and enhance the environment. While sustainable development is essentially a global concept, green growth is much more feasible at a local scale.

Any eco-industrial revolution will need to:

- maximise efficiency in the use of environmental resources and energy, using technology to reduce use of resources and production of wastes (known as green, clean, lean technology)
- redirect corporate energy to satisfy broader human needs and to ensure that the world's people enjoy sustainable livelihoods — a particular issue for the growing number of workers in developing countries

This revolution requires different priorities and different thinking by organisations and companies. However, support for green growth has built up steadily over the last 30 years, not least because it can be seen as profitable — as shown by 3M's pollution prevention strategy (developed in the 1980s). The strategy did not decrease profits — it actually enhanced business via a more positive company image. Many global companies now publish environmental reports and realise that good environmental practice is actually good for business.

GENERAL ELECTRIC: THE BUSINESS OF GREENER GROWTH

Case study **15**

Many global companies are putting environmental issues central to their business agenda. One example is General Electric (GE), the world's largest company (see Table 5, p. 18).

Over the 5 years from 2004, GE will:

- double its R&D spending, by investing $1.5 billion annually into research on cleaner technologies

- double its revenues from: renewable sources of energy; technologies and materials that make energy consumption more efficient (e.g. better aero engines); cleaner and more efficient transportation technologies; and products and services that conserve or purify water
- reduce greenhouse gas emissions by 30%, and improve energy efficiency by 1%, by 2012; if the company followed a 'business as usual' policy, emissions would increase by 40%

This policy change was sold to managers and workforce, who previously had a profit-making culture. GE's commitment has to be seen as green growth — strong economic performance but with less environmental damage. As environmental monitoring and legislation develop, GE's strategy may be seen as enlightened future-proofing.

Website: **www.ge.ecomagination.com**

Green growth has been around for longer in the service sector, as the following major case study shows.

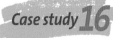 *Case study 16*

ECOTOURISM IN NEW ZEALAND

Economies are rarely sustainable at a regional scale. However, there are many small-scale initiatives at the local scale that attempt to be sustainable while at the same time promoting green growth in an economy. This case study examines the impact of ecotourism at both a local and a regional scale on a small area of South Island in New Zealand.

Ecotourism is 'responsible travel to natural areas that conserves the environment and sustains the wellbeing of local people' (International Ecotourism Society).

The west coast of the South Island of New Zealand was a relatively remote and unopened wilderness well into the twentieth century. Although Maoris and, later, non-Maoris travelled the coast for centuries, the Southern Alps were almost impassable from the east due to the steepness of the terrain and so the west coast remained cut off. Much of the coastal region is part of the South West New Zealand World Heritage Area (Te Wahipounamu), established in 1986 and enlarged in 1990 to include Westland, Aoraki/Mount Cook, Fjordland and Mount Aspiring National Parks and several wilderness areas. The heritage area in the southwest of New Zealand covers 2.7 million hectares (10% of the country) and has been little modified by human influences other than at a few locations. The area contains the world's finest remaining examples of Gondwanaland flora and fauna in their natural habitats.

The west coast can be divided into three distinct areas:
- the northwest, beyond the heritage area, where the ranges of the Southern Alps are further inland
- the area extending southwest from Hokitika to Haast, including the 127 500 ha Tai Poutini National Park
- the Fjordland region of the World Heritage Area

It was gold mining that first brought people to the coast of the northwest area in large numbers. The land has since been used for agriculture and some mining of coal. This area stretches from Westport southwest to Hokitika and is north of the study area.

The area extending southwest straddles the Alpine Fault plate boundary. East of the Alpine Fault are the forested Southern Alps, rising steeply with slopes cut by impassable

Contemporary Case Studies

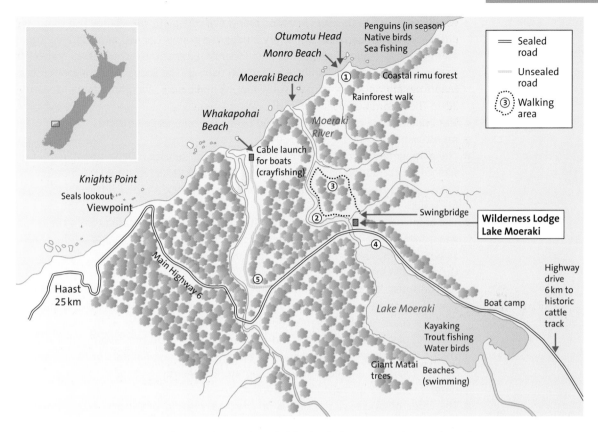

Figure 46
The location of
Wilderness Lodge,
Lake Moeraki

gorges, to permanent snowfields. These snowfields feed the Fox and Franz Josef glaciers, which descend right down to the lowlands. Dense temperate rainforest covers the lowlands west of the Alpine Fault, with lakes, wetlands and wide river mouths. This central section has only been opened up in the past 40 years since the completion of the road from Fox south to Haast and inland to Wanaka in 1969. This is the section where Lake Moeraki Wilderness Lodge is located (Figure 46).

The Fjordland region is an alpine wilderness dissected by fjords such as Milford Sound and Doubtful Sound.

Lake Moeraki Wilderness Lodge

Until 1965, Lake Moeraki was only reached by a 90-km gravel road from the north, which had to ford many rivers. It took 4 years to construct a 30-km gravel road from Lake Moeraki to Haast because of the difficult coastal terrain. A camp grew up at Lake Moeraki to house the men working on the road, which then became a motel for people travelling the road. The road has since been surfaced and is now Highway 6. In 1989, the motel was acquired by two ecologists who redeveloped the site into the 28-room Wilderness Lodge. Their aim was to help protect the rainforests and to introduce visitors to untouched surroundings. In 2003, the Wilderness Lodge was rated as one of the top 20 hotels in the world by *Fodor's Guide* on the basis of its setting, activities, ambience and service as well as its location and construction.

How green and sustainable is Wilderness Lodge?
■ It has a commitment to environmental protection, which is made clear to guests.
■ It runs its own hydroelectric generator, which generates 35 kW of power for the

whole lodge. A fish trap was installed below the HEP plant to help any fish that have difficulty swimming upstream. Every 2 days, the trap is opened and fish are carried above the hydro and released into Lake Moeraki. Environmental scientists state that the power station has very little impact on fishing. The Lodge works with the local Maori people, Te Runanga o Makaawhio, to make sure that the scheme does not harm their source of fish.

- The buildings are energy-efficient and usage is also energy-efficient.
- All rubbish is recycled.
- The management works with the New Zealand Conservation Department, gathering scientific information on fauna and flora. The lodge has a registered seismometer to measure the frequent earthquakes along the complex Alpine Fault separating the Australian Plate from the Pacific Plate.
- The lodge uses wilderness food — whitebait from the river, seafood from the coast, game from the forest and some own-grown vegetables. New Zealand produce, such as wines from Marlborough and Otago, are used.
- Guided walks are organised with qualified ecologists to observe penguin nests in the coastal forest, seals on Monroe Beach, Hector's dolphins, freshwater crayfish and giant eels, and the unique local flora such as podocarp and rimu trees. The forest contains more 'perching plants' (epiphytes) than anywhere else in the country.
- Invading pests, such as ferrets and possums, are killed to preserve the native food chains. The Moeraki River headwaters are one of the last habitats of the blue duck — the focus of a special rescue programme. Nearby is another special rescue programme for the rarest kiwi, the Tokoeka.
- The lodge organises canoe safaris on the lake with no motorised water transport. Trout and salmon fishing is possible on the lake and river.
- It is a small-scale enterprise.
- Guides to the local environments are sold to educate guests. However, most guests are already environmentally conscious.

Wilderness Lodge, Lake Moeraki

Contemporary Case Studies

- The lodge has its own wastewater treatment.
- It is owned and operated by New Zealanders. It is not part of a chain returning profits beyond the country and it provides jobs in a region of few opportunities.
- There is no other habitation within 40 km.
- The owner is President of the Royal Forest and Bird Protection Society. He has successfully campaigned to stop the logging of the ancient podocarp trees in the area, which are often up to 800 years old.

22

Using case studies

Question

Using the 'sustainability quadrant' below as a framework, assess how sustainable Wilderness Lodge is.

Futurity	Eco-friendliness
Equity help of the poorest people	Bottom up, involving local people

Where greenness 'fails'

- Guests arrive by car. They are mainly from North America and Europe, the two largest source regions of ecotourists. Air travel is a major source of greenhouse gases.
- Some of the organised walks to study the environment need transport to and from the locations.
- The local HEP station is not always enough for peak use, so the lodge has its own generator.
- Food miles — bringing produce to this remote location uses energy.
- Ecological change due to human influence is prevented by killing pests, making the ecosystem fossilise rather than change as a result of the invasion of pests.
- Some money trickles out to pay commissions to travel agencies.
- The price of accommodation at the lodge is high and so only wealthy people tend to stay.

Can such a single development contribute to a sustainable region?

The Department of Conservation (DOC) in New Zealand has looked at the impact on the national economy of 1.9 million hectares of conserved land on the west coast (which includes Lake Moeraki Lodge). This peripheral region has experienced declining employment in forestry, mining and industry since 1986, whereas all the employment categories associated with tourism (restaurants, accommodation, cultural services and retailing) have grown. The region is a major source of tree mosses for horticulture and floristry.

Figure 47 shows that over 40% of visitors to Wilderness Lodge come for nature, and for bush, coast and sea. 75% have come to enjoy the conserved environment. The activities that attract over 50% of the visitors are associated

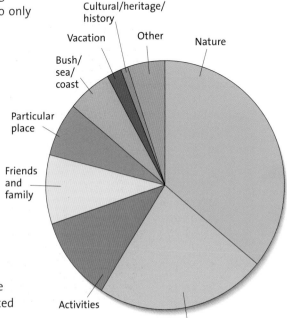

Figure 47 Why visit the west coast wilderness?

Cultural/heritage/history
Vacation
Other
Nature
Bush/sea/coast
Particular place
Friends and family
Activities
Glaciers

with the unspoilt bush, seashore and the wild coast, all of which are within minutes of Lake Moeraki Lodge.

Visitors bring NZ$209 million (approximately £40 million) into the area and have generated 1870 jobs on strictly controlled concessions in the conservation area. It seems that green tourism in such a conserved environment does make a significant contribution to the region's income, which the DOC estimates at 12% of the value added by all economic activities. Green tourism managed sustainably through concessions to run any activity within the area (such as Wilderness Lodge) is seen as sustainable and of increasing importance as other resources, such as mining, decline.

Websites: **www.wildernesslodge.co.nz**
www.doc.govt.nz
www.tourismconcern.org.uk

Green tourism is growing faster than conventional tourism, especially in areas where there is a large number of protected areas. In *Case study 16*, the New Zealand government pays for upkeep, whereas in Costa Rica, users are expected to pay for upkeep. Ecotourism can easily degenerate into mass tourism — a victim of its own success. This is a perceived problem, both in Costa Rica and the Galapagos.

Many governments, businesses and other decision makers are arguing for green growth as the only way ahead. Satellite images showing the destruction rate of the world's ecosystems and the melting of the Arctic ice sheet, combined with concerns over supplies in the world's key energy and mineral resources, make green growth vital. The Chinese government adopted green growth as a key component of its massive industrialisation programme in 2005, as a result of some extreme pollution incidents. Equally, increasing numbers of transnational and global companies set sustainability as a vital component in their future growth. With the size of green consumerism, many large companies and governments see it as a marketing tool too.

However, the path to global sustainable economic development is fraught with difficulties and conflict. Working incrementally from a local scale to a regional scale seems the way ahead, even with all the difficulties. National governments, both individually and collectively at world forums such as Johannesburg 2002 (the 'Sustainable Summit'), can provide the legislative framework and environmental regulation and monitoring to make things happen.

23 Using case studies

Question
Outline the economic and environmental costs and benefits of one green tourism scheme.

Guidance
Remember, the question asks for costs *and* benefits for the economy and the environment. These can be local, regional or national.

Examination advice

The case studies in this book, together with others that you have studied and learned, are an integral part of your A-level geography course. They can be put to good use in a variety of ways in the examination. Much depends on the task in the question and the command words used. Table 29 illustrates a variety of question scenarios.

Table 29 *Question commands and the examiner's expectations of case studies*

Task or command*	Required case study detail			
	Given data	Support	Compare	Evaluate or assess
Typical question	Study the map showing the hierarchy of global economic centres. Describe the pattern and suggest reasons for it.	Examine the impacts of global shifts in manufacturing on the environment and economy of one NIC.	With reference to a range of examples, discuss the factors that are assisting the process of globalisation.	Assess some of the ways in which Third World debt may be reduced.
Case study expectation	You need to refer to all the levels on the map, but you may give reasons that are based on one example of each of the levels in the hierarchy.	Give detailed examples that name regions, towns and cities in your chosen country.	More than one country and more than one industry are required to provide the required range.	Give examples of different approaches to debt relief, naming the origin of the policy and examples of where it has succeeded and failed.

*A word of caution — some questions may not ask for examples, but nonetheless the examiner will be expecting them. For example:

Discuss one of the following statements:
- New technologies have had limited impact on lifestyles.
- Restructuring debt in LEDCs has not been very successful.
- Green growth is frequently not viable.

If you are in doubt as to whether examples are required, it is best to give some.

Using stimuli provided in the examination

So much depends here on whether you will be given marks for describing the resource, as shown below, or whether the question just says 'suggest reasons' for the trends or patterns shown. In the latter case, it pays to annotate the diagram to help you provide a structure, and then to include evidence from the resource to support your answer as there are no marks for description.

Many questions at A-level involve the use of maps and diagrams to test your knowledge and to give clues to the required answer.

Question

The graph in Figure 48 shows the top ten countries that rely heavily on the export of IT products.
(a) Describe the spatial pattern shown on the graph.
(b) Suggest reasons why exports of high-technology products are important to countries such as those in Figure 48.

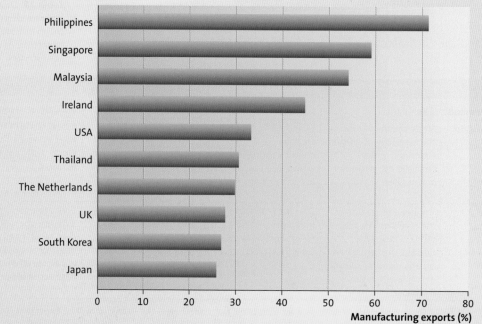

Figure 48 *IT products as a percentage of total manufacturing exports*

Guidance

(a) The graph shows a mixture of NICs in the Far East and southeast Asia and developed countries where the technology originated. Always quote precise data.

(b) This is where you can use the stimulus of the graph or select another similar country (e.g. Taiwan). Make sure that you know why IT is important to at least one NIC and one MEDC. Examples of TNCs, the R&D base of companies such as Intel and labour costs in NICs would help.

Using more than one case study

Sometimes examiners give you the option of selecting your own examples. Do pay attention to whether the important word in the question is the singular, or, as in the following question, the plural.

Contemporary Case Studies

Question

Using examples, discuss how governments attempt to influence the location of industry.

Guidance

Ideally, your examples should be taken from at least two countries in different parts of the world, although, in this question, it might be possible to use the UK and Italy. Often the question asks for a *range* or for contrasts, so be prepared to justify your choice in the introduction to your answer.

Here are some ideas:

- government policy towards the location of the steel industry in the UK
- policies towards the location of automobile assembly plants, for example VW in Emden, Nissan in Sunderland, or Ford and Vauxhall on Merseyside
- the role of Malaysian and other NIC governments in the development of IT-based industries

Plan the answer carefully, using a numbered order or a diagram to make certain that you cover a range of points

Building an essay around a single case study in depth

A successful response to any question that starts 'With reference to a named...' requires you to:

- choose an appropriate country or company
- have a sufficiently detailed knowledge of the country or company to provide the points that the examiner expects
- avoid the 'in Taiwan...' or 'my chosen company is ABC Electronics...' approach, if there is then not a single reference to any specific location in the country, or you simply give a general essay on electronics without further reference to the named company
- avoid the approach that puts down all you recall from a homework question that you did during your course but does not actually provide an answer to the question that you have been asked

Question

With reference to one or more named TNC, examine its role as a global producer and employer.

Guidance

For a more complex question such as this, it is always advisable to deconstruct it.

- Your introduction should make it clear whether you are going to use a study of a single company (e.g. VW) or look at more than one TNC (e.g. Mittal *and* Intel).
- Note that the question wants to know about *production* — i.e. where and why.
- The second part asks about *employment*. You might not have details, but do find out where the majority are employed and perhaps whether R&D is in an MEDC whereas other plants are merely assembly lines or workbench sites.

Questions involving evaluation and assessment

Here data are provided, mainly to act as a stimulus for your knowledge. *Assessment* requires you to say why some implications are more important than others.

Note that you do not have to use the examples given in the question — for instance, you could use bananas and St Lucia, if these are your known case studies, but you do need to identify some implications from the tables before you start (as shown) and then apply these implications to your chosen case study. Avoid a merely descriptive use of your case study — the 'All I know about bananas in St Lucia…' approach.

Question

Assess the implications of the trade data for the economies of the least developed countries (Tables 30 and 31). You may refer to other commodities and countries.

Country	Main commodity	Main commodity as % of exports
Uganda	Coffee	56
Zambia	Copper	56
Mali	Cotton	46
Rwanda	Coffee	45
Chad	Cotton	42
Burkina Faso	Cotton	39
Benin	Cotton	38
Guyana	Gold	16
Tanzania	Coffee	11

Table 30 *Commodity dependence*

Country	1998	1999	2000
Uganda	−5	−17	−34
Zambia	−20	−26	−28
Mali	−11	−23	−28
Rwanda	6	−11	−25
Chad	−6	−15	−20
Burkina Faso	−4	−16	−25
Benin	−7	−14	−16
Guyana	0	7	−14
Tanzania	1	−7	−13

Change in terms of trade is the average price change of the three main export commodities relative to their price in 1995–97.

Table 31 *Change in terms of trade*

Notes for Table 30:
- there are issues of over-dependence
- general dominant nature of main commodity in exports — vulnerability from diseases, price fluctuations etc. and competition

Notes for Table 31:
- note the escalation of decline in all cases
- there are implications for farmers, trade balance etc.

Guidance

- Single crops may fail due to drought or conflicts. New, cheaper competitors may arrive, for example coffee from Vietnam. TNCs are forcing the prices down, for example by using plantations rather than small-scale producers.
- A decline in prices obtained means less income for the country and the people. The country is then unable to invest tax revenue in education and health. Therefore, development is retarded or reversed.
- Which implication is the most threatening to the countries, or do different implications have greater importance in some countries?

Evaluation, with selection of supporting examples

Again, this question leaves the choice of examples open to you. However, it does expect you to give some indication of the most important examples.

28 *Using case studies*

Question
Evaluate the roles of trade, aid and debt relief in promoting economic development in the least developed countries.

Guidance
- Make sure that you select LDCs (see Part 1).
- Ideally, choosing a different country for each phenomenon might be best, because it enables you to suggest that it depends on the existing level of development of the country.
- On the other hand, you may have details on the past misappropriation of aid, the problems of trade and proposed debt relief conditions for one country, such as Uganda.
- Equally, you could adopt a wide ranging approach, but remember to avoid giving examples that start with 'in Africa…'. Africa is a *continent*, while the question asks about *countries*.
- Try to write three sentences to support your example, using facts and figures.

Responding to a cartoon stimulus

The cartoon in Figure 49 is meant to stimulate your ideas and leave you to develop your own case studies.

Figure 49

Ingram Pinn (*Financial Times*, 19 January 2005)

Question

To what extent is the cartoon an oversimplification of the problems of aid to HIPCs?

Guidance

- Try to work out what the cartoon is 'saying' and then state this in your introduction.
- Make sure you select HIPCs (one, two or more, but all must be HIPCs).
- Your case studies should look at countries where aid has been affected by war, such as the Sudan and Ethiopia.
- Contrast this with countries where aid is swallowed in bureaucracy and corruption (Nigeria, perhaps).
- Also look at countries where aid is hampered by environmental factors, such as seasonal rains (Niger).
- In all cases, the aid must fail to help the poorest peoples who need it so badly.

Index